GRAVITY OF COMMAND

LEADING WITH PURPOSE IN THE AGE OF ARTIFICIAL INTELLIGWNCE

By

Shailesh Shrivastava. And

Taylan Yildirim

The Gravity of Command[i]

ISBN: 979-8-9953078-0-8

Dedication

To all mid-level managers who quietly hold chaotic project togethers and bravely defy project gravity. To those who dodge the black holes disguised as deadlines, navigate the cosmic scope creep, and pilot their teams toward the promised land of "on time" and "on budget," your leadership keeps our world spinning through the shifting priorities, ever-changing customer requirements, and all the AI buzzwords.

About The Authors

Taylan Yildirim, Co-Author & Leadership Strategist

Taylan Yildirim is a Seattle-based executive leader, speaker, and the author of *The Gravity of Command: Leading with Purpose in the Age of AI*. With over 25 years of experience at the forefront of telecommunications and digital transformation, Taylan specializes in navigating the high-stakes intersection of ethics, disruptive innovation, and organizational purpose.

Taylan has spent his career operating at the intersection of complexity and consequence — where technology, scale, and leadership collide. From steering large-scale telecom transformations to navigating demanding operational environments, he has built a proven reputation for turning ambiguity into execution. Over the past two decades, he has led global teams of more than a thousand engineers, operators, and problem-solvers, delivering mission-critical programs across cloud transformation, automation, and AI-driven systems. His work lives not in theory, but in resilient systems that perform — every day, at scale, without failure.

Alongside his executive role, Taylan is currently pursuing a doctorate in business management, with research focused on how artificial intelligence is reshaping governance, decision-making, and the architecture of modern organizations.

Shailesh Shrivastava, Co-Author & Strategic Advisor

Recognized as "The Strategy Guy," Shailesh Shrivastava is a visionary technology executive and seasoned business leader with over two decades of experience at the epicenter of enterprise transformation and emerging technology. A hands-on enthusiast and early adopter, Shailesh has dedicated his career to decoding the complex interplay between data, advanced systems, and human behavior to empower premier tech companies through their most critical business decisions. From his foundational days as a technical engineer to directing global technology sales and strategic development in the telecommunications sector, Shailesh delivers a rare, "full-stack" perspective to leadership.

His contributions to *The Gravity of Command* illuminate a vital truth: while AI will automate decision-making, the executives who thrive will be those who anchor their organizations with an unwavering sense of purpose and authentic human connection.

A principal architect of TM Forum's customer experience and strategic frameworks, Shailesh translates complex digital transformation into an actionable "survival guide" essential for the AI era.

Introduction

"Leadership wasn't always rocket science," Bob says, adjusting his glasses. "But these days, with AI in the cockpit, every decision feels like calculating for re-entry." Welcome to The Gravity of Command, your practical field guide for leading at the speed of technological change.

We stand at a unique precipice in the history of project and program management. The landscape of information technology is not merely evolving; it is undergoing a seismic transformation driven by the relentless advance of artificial intelligence. For those of us who have spent decades managing teams, navigating critical paths, and delivering on time and on budget, the familiar tenets of our profession are being challenged, augmented, and fundamentally rewritten.

In today's world, where automation, predictive analytics, and digital acceleration are everyday realities, even experienced leaders recognize that the old map no longer fits the new territory. We are seeing a shift from the direct oversight of predictable processes to the strategic orchestration of adaptive, intelligent systems.

The New Gravity

Leadership has always been about more than titles or corner offices. It is the subtle force that holds teams and missions together, binds people to a purpose, and prevents operations from spinning out of control when the pressure mounts. We call this Gravity Leadership.

Just as gravity gives planets their orbits, your leadership is the invisible force that keeps your project anchored when deadlines loom and scopes creep. But today, AI is changing the very spin of your solar system. It acts as a "gravity booster," amplifying your team's reach, automating the "black holes" of repetitive work, and uncovering insights once hidden in the cosmic dust of your data.

Bridging the Gap: Bob and Alice

To navigate this new frontier, we must bridge the gap between established leadership principles and emergent realities of AI. Throughout this book, you will meet two archetypal leaders who represent this duality:

- **Bob:** The veteran manager and master of purposeful leadership. He excels at team connection and navigating the human elements of project delivery, but finds himself challenged by the rapid shifts of the AI era.
- **Alice:** The new-age AI leader. Sharp and data-savvy, she navigates algorithms and automation with confidence but grapples with the deeper demands of vision, purpose, and ethical balance.

Whether you identify more with Bob's experience or Alice's technical fluency, this book is designed for you: the project orchestrator, the team lead, and the modern manager navigating uncertainty.

Your Mission

We are no longer just managing systems; we are orchestrating adaptive, intelligent processes. As artificial intelligence weaves itself into the fabric of our work, from more intelligent system alerts to autonomous operations, your most important asset is not just technical skill but the ability to connect people around a shared purpose.

The Gravity of Command is a recognition of the weight of leadership and a call to embrace this responsibility. It is time to stop reacting to the future and start shaping it. Let's suit up and take command.

Front Matter

This book is dedicated to the intrepid professionals who stand at the helm of technological endeavors, navigating the swirling currents of Artificial Intelligence and the gravitational pull of responsibility.

To all mid-level managers who bravely defy project gravity. To those who dodge the black holes disguised as deadlines, navigate the cosmic scope creep, and pilot their teams toward the promised land of "on time" and "on budget," your leadership keeps our world spinning, our code compiling, and our teams safely in orbit around fast-moving deliverables.

This work is a tribute to your dedication, your resilience, and your unwavering commitment to driving progress. May you find in these pages the guidance and inspiration to lead with purpose, to embrace the challenges of artificial intelligence with confidence, and to truly understand and wield the profound "Gravity of Command" in your roles. Your leadership is the unseen force that transforms ideas into reality, and this book is a testament to your vital contribution.

Prologue

The landscape of information technology is not merely evolving; it is undergoing a seismic transformation, driven by the relentless advance of artificial intelligence. Consider the scale: Gartner forecasts $2.52 trillion in worldwide AI spending for 2026—a 44% year-over-year surge. Yet here is the paradox that defines your leadership challenge: McKinsey[1] estimates generative AI could add $4.4 trillion annually to global GDP (roughly 4% of worldwide output), yet only 1% of organizations describe their AI deployments as "mature." The gap between potential and reality is where Gravity Leadership must anchor your teams.

This book emerges from urgent necessity. Not because AI is coming—it is already here. But because 95% of enterprise GenAI pilots fail to deliver ROI within six months, and 60% of AI projects are abandoned by the end of 2026 due to poor data readiness.

For those of us managing teams and projects in this dynamic sphere, the familiar doctrines of project and program management are being challenged, augmented, and sometimes fundamentally rewritten.

This book, *The Gravity of Command*, is to bridge the gap between established leadership principles and the emergent realities of an AI-integrated world. It is intended as your practical companion, your mentor in print, for mid-level IT and tech professionals who aspire to lead not just with competence, but with purpose and foresight. We are no longer only managing systems; we are orchestrating adaptive, intelligent processes. The integration of AI into our daily workflows, from automating repetitive tasks to enhancing decision-making, demands a renewed understanding of leadership.

To guide you on this journey, each chapter features themed sidebars to keep your thoughts organized and your momentum strong:

---

[1] *McKinsey Generative AI Report (Jan 2026)*

Gravity Pulls	Key lessons and takeaways that distill the strong force of leadership and anchor your project, and keep your team orbiting together.
Light Year Lessons	long-term perspective and wisdom, helping you expand your leadership horizon and stay ahead for the future.
Mars Rovers	Provides practical steps, efficient tactics, and smart reminders you can deploy instantly when navigating uncertain terrain

- **Orbiting Truths:** Core principles and foundational insights to keep your leadership grounded.
- **Mission Control:** Practical steps and actionable strategies to implement immediately in your projects.
- **Warp Speed Tips:** Quick hacks and efficient methods for navigating the fast-paced tech environment.
- **Black Hole Alerts:** Warnings and common pitfalls to avoid when integrating new technologies or leading teams.

Through real-world examples and hands-on lessons, you will learn how to navigate team dynamics, project complexities, and program oversight in an era where AI is not a distant possibility but an everyday reality. The "Gravity of Command" is a recognition of the weight of leadership, a call to embrace this responsibility, and a toolkit for shaping the future of technology.

Whether you are charting a steady course or skipping across chapters at warp speed, these sidebars will serve as your cosmic signposts for quick learning, practical action, and lasting wisdom.

Table of Contents

1. The Sun of Purposeful Leadership

Gravity: The Invisible Force

"People don't follow titles. They follow purpose." - Bob

Historically, the management style was often characterized by a directive approach, with technical direction flowing from the top down and team members expected to execute tasks with precision. This model, while effective in its era, is increasingly insufficient in today's fast-paced, AI-driven world. The sheer pace of technological change, coupled with the transformative power of AI, demands a leader who can not only keep pace but also anticipate and shape the future.

The Migration Crisis: A Tale of Two Leaders

Let's bring this idea to life by returning to our two guides: Bob, the "Boss of Back Times" and top performer of 2020, representing the tried-and-true, classically effective manager; and Alice, the New Age AI Leader of 2026, who navigates a whole galaxy of new challenges.

Meet Bob & Alice

Bob

Bob, a master of best practices and purpose-driven values, struggles to keep pace with the demands of AI, including advanced analytics and real-time decision-making.

Alice

Alice, a technically skilled leader, excels in AI but needs to integrate emotional intelligence and ethical clarity into her approach.

Bob's sharp, intelligent eyes had witnessed countless technological evolutions. He has been a legend in the company, the architect of processes that have guided them through decades of success. His approach was steeped in best practices, a deep understanding of purpose-driven values that resonated with his teams. He could dissect a complex problem, chart a course, and inspire dedication through sheer clarity of vision. Yet, lately, a subtle unease had begun to creep in. The sheer velocity of AI, the blinding speed of advanced analytics, and the constant demand for real-time decisions felt like shifting sands beneath his well-worn boots. He was a master of the old ways, but the new language of AI, with its predictive models and emergent capabilities, was a dialect he was still struggling to comprehend fully.

Alice, on the other hand, was a creature of this new AI-driven era. Her late-thirties frame was topped by a cascade of curly hair, framing a face quick to break into a confident, almost mischievous smile. Her eyes, bright and keen, danced with the exhilaration of understanding complex algorithms and data streams. She could coax insights from raw information with an almost intuitive grace, her technical prowess undeniable. She was a testament to the future, a leader who thrived in the heart of AI's transformative power. However, as she navigated this new landscape, a different kind of challenge emerged. The data whispered truths, but the human heart often screamed questions. Her journey, as she was beginning to realize, was not just about mastering algorithms; it was also about weaving empathy and ethical clarity into the very fabric of her leadership.

The two stood, in a sense, at opposite ends of a technological spectrum, yet both were grappling with the fundamental truth: People don't follow titles. They follow purpose. Bob's purpose had been forged in the fires of established order, a clear beacon guiding his teams. Alice's purpose was still unfurling, a vibrant tapestry woven from innovation and the potential for a better future, but one that now required a more nuanced understanding of human connection and ethical responsibility. Their very presence in the same room, at this pivotal moment for the organization, was a testament to the evolving nature of leadership itself.

They were sitting in the war room, recounting a massive database migration project from the previous spring.

“Remember that migration?” Bob asked, cleaning his glasses. “The database system was like an eclipse on a full moon night, and it decided to keep us in the dark right after the upgrade. Deadlines were circling like asteroids, and the team chat looked like mission control during a failed launch.”

Alice nodded. She remembered the panic. The automated alerts were firing so quickly that the team couldn't read them.

“Honestly,” Bob continued, “what got us through wasn’t just the technical patches. It was keeping everyone’s focus on why we were launching in the first place. We rallied not around debugging code, but around the promise of delivering real-time value to our customers. That reminder kept us in orbit.”

Alice smiled. “I used to think the right algorithm could solve anything. But when things go dark, even a machine needs a North Star. Purpose gave us gravity.”

The North Star: Setting a Guiding Vision

Gravity keeps the team together, but gravity alone doesn’t tell you where to move. It just keeps you from drifting apart. To move with velocity, you need a vector. You need a North Star.

In the age of AI, we are often overwhelmed by data, metrics, and algorithmic predictions. We have more maps than ever before, but fewer directions. This is where the leader must step in to define the ultimate destination. Without a North Star, leadership becomes guesswork, like driving through the Texas plains on a moonless night, hoping you’ll spot a landmark before you run out of gas.

For Alice, the North Star is the "Strategic Intent." It is the filter she uses to judge the thousands of suggestions generated by her AI tools.

- The AI might say: "Optimizing this server cluster will save $400 a month."
- The North Star says: "Our mission is Maximum Reliability for the patient data. We do not compromise reliability for small savings."

Without that North Star, Alice might have approved the cost-cutting measure, potentially risking the entire project's integrity. The North Star turns "data-driven decisions" into "purpose-driven decisions."

Gravity Leadership

Leadership. Some people wear it like a shiny lapel pin of authority—something you earn when someone hands you a title, a desk with a view, and the power to say, "Let's circle back on that."

But real leadership? It's not about badges, power moves, or TED Talk sound bites. Leadership is something deeper. It's the subtle force holding teams and missions together—binding people with purpose, pulling others into your orbit, and quietly preventing the whole operation from spinning out of control when a meteor shower (or, let's be honest, a failed deployment) strikes.

Imagine leadership as gravity itself. You cannot see it, but you feel it everywhere. It's what keeps teams grounded when deadlines loom, budgets shrink faster than your weekend, or your API services go down after an upgrade. The gravity of leadership keeps ideas from floating off into space, keeps accountability from evaporating in the virtual or remote work vacuum, and keeps your project anchored to something real when the latest buzzword threatens to flatten your sanity. Without this force, people drift.

And this is not just a metaphor.

Research involving more than 400 professionals found that leadership behaviors that build trust and team cohesion significantly increase job satisfaction and team stability[ii]. In that study, leadership factors explained over half of employee satisfaction outcomes. Even more telling, in several cases, leadership influence only worked when trust was present. Trust acted as the transmission mechanism between leadership behavior and team morale. In simple terms, without trust, influence weakens.

For a new manager, this changes everything. Your authority does not create engagement. Your credibility does. When your team members trust you and feel connected, performance stabilizes. Engagement rises. Resilience increases. Trust and cohesion are not "soft" qualities—they are structural forces. Many modern managers fear that AI will dilute their influence. The reality is the opposite. As artificial intelligence rises to automate tasks, the true gravitational force of leadership becomes not less important, but essential. AI acts as a "gravity booster." It amplifies your team's reach and automates the black holes of repetitive work. But amplification without direction only increases chaos.

We are moving from managing systems to orchestrating adaptive, intelligent processes. Your real power lies in knowing why you do what you do—and in creating a center of gravity strong enough to hold your team together when the algorithms shift. As artificial intelligence automates tasks and analyzes data at lightning speed, the uniquely human elements of leadership become even more critical. Machines may handle the numbers, but they cannot inspire, unite, or guide teams through uncertainty.

That remains your domain.

Gravity Pulls	*Artificial intelligence acts as a "gravity booster," amplifying a leader's reach and automating repetitive work, but it makes the human aspects of leadership—such as inspiring and guiding teams—more essential.*

[Sidebar] Gravity Pulls: The Core Principles

- **The Invisible Anchor:** Leadership is the force that prevents "drift" in remote and automated environments.
- **The North Star:** Data gives you options; the North Star gives you the decision. Without it, you are just reacting to metrics.
- **Purpose Over Compliance:** In the AI era, you cannot manage by directive alone. You must align energy and vision so the team orbits the mission voluntarily.

Gravity leadership is not just about delivering impact but aligning energy, vision, and values so your team orbits something far bigger than any one of them could imagine alone. You set a North Star so magnetic that even skeptics quietly grab their telescopes.

Small Actions, Big Impact

A common misconception among new managers is that purpose requires grand speeches or sweeping strategic pivots. In reality, gravity is often generated by the smallest objects.

Researchers who study failure in complex systems have long observed a simple truth: catastrophes rarely begin with dramatic collapse. They begin with small, overlooked deviations.[iii] In highly interconnected environments, minor issues can cascade across tightly coupled systems, producing outsized consequences. This is the core idea behind Normal Accident Theory, introduced by sociologist Charles Perrow. His argument was uncomfortable but powerful: in complex, tightly interconnected systems, accidents are not rare exceptions—they are structurally inevitable. When components are deeply interdependent, small errors do not stay small. They spread.

Consider the story of an Italian automobile company known for its precision engineering. When vehicles began failing unexpectedly, leaders suspected advanced electronics or design flaws. The culprit turned out to be a single loose nut in the steering column. One overlooked component destabilized the entire system. This aligns with another influential framework, the Swiss Cheese Model[iv]. Imagine multiple layers of defense: policies, procedures, technology, and oversight. Each layer has small holes. Normally, those holes do not align. But when they do, failure passes through every layer. Disasters are rarely caused by one big mistake. They occur when several small vulnerabilities line up at the same time.

The Titanic was a marvel of engineering. Yet ignored warnings, missing binoculars, and underestimated environmental conditions aligned. In modern project environments, the equivalent may be a vague requirement in a user story, a skipped stand-up, an unchallenged assumption in an AI model, or an unanswered stakeholder email. Individually trivial. Systemically dangerous.

But there is another side to this research.

While Normal Accident Theory highlights structural vulnerability, High-Reliability Theory offers a counterpoint. Researchers studying nuclear submarines, air traffic control systems, and aircraft carriers found that some organizations operate for decades in high-risk environments with almost no catastrophic failure.[v]

These organizations are called High-Reliability Organizations (HROs).

What makes them different?

They assume failure is always possible. They stay preoccupied with small anomalies. They treat near misses as learning opportunities. They continuously ask, “What are we overlooking?”

This mindset—what researchers call a preoccupation with failure—is not pessimism. It is disciplined attention.

Gravity Leaders adopt the same posture.

They do two essential things.

1. Recognize the Invisible Work

High-reliability research shows that heroic interventions do not sustain safety and performance. They are sustained by the disciplined execution of routine work [vi].

In aviation and healthcare, near-miss reporting systems exist precisely because small deviations matter. A nurse reporting a labeling inconsistency may prevent a fatal dosage error later. A technician flagging a vibration anomaly may prevent equipment failure.

Think of hospital cleaning staff. Their daily routines reduce infection rates and save lives. Without them, clinical excellence collapses.

In technology organizations, invisible work looks like:

- A junior engineer writing regression tests.
- An admin verifying access rights.
- A data analyst checking edge cases in a training set.
- An operations lead reviewing minor incident logs.

None of these actions generates headlines. But they close the holes in the cheese.

Gravity Leaders elevate this work. They connect it to the mission. They protect it from being deprioritized.

2. Connect the "Why" to the "What."

Another concept from High-Reliability research is collective mindfulness[vii]. This means people understand how their local actions affect the broader system.

In tightly coupled environments, local optimization can create global instability. If a team member improves one metric without seeing systemwide consequences, risk increases.

Imagine asking someone to temporarily manage ID access requests. Framed narrowly, it feels administrative. Framed strategically — "This ensures new hires are productive on Day 1 and protects system security during onboarding" — the task gains gravity.

When people understand systemwide impact, compliance becomes commitment.

The New Frontier: From Overseer to Orchestrator

For leaders like Alice, the challenge is even greater.

Modern AI-enabled systems are:

- Complex (many interacting components),
- Tightly coupled (changes propagate quickly),
- Dynamic (conditions evolve in real time),
- And probabilistic (outputs are not deterministic).

In these environments, small data biases can scale globally. Minor integration gaps can create cascading failures.

Normal Accident Theory would caution: the more complex and interconnected the system, the higher the structural risk.[viii]

High-Reliability Theory would respond that risk can be significantly reduced, but only through disciplined organizing, continuous vigilance, and strong cultural norms.[ix]

The leadership shift is profound.

The Old Way (Bob's Era):

- Perfect the process.
- Enforce compliance.
- Optimize execution.

The New Way (Alice's Era):

- Orchestrate human-AI collaboration.
- Encourage anomaly reporting.
- Maintain psychological safety so concerns surface early.
- Anchor innovation to business value.

High-reliability organizations balance two forces:

- Standardization (to reduce variation),
- Adaptability (to respond to the unexpected).

Too rigid, and the system cannot react.

Too flexible, and discipline erodes.

Gravity Leadership lives in that tension.

It is not about being the heaviest object in the room.

It is about creating a stable orbit — where small details are respected, invisible work is honored, and purpose connects local effort to systemwide impact.

Because in complex systems, small actions compound.

And gravity is built one decision at a time.

*****

Checklist: Are You Generating Gravity?

- [] **The "Why" Check:** In your next team meeting, can you explain why the current sprint matters to the company's five-year goal in one sentence?
- [] Spot the Invisible: Identify one person on your team doing "invisible work" (maintenance, documentation, QA) and publicly recognize their impact on the project's stability.
- [] **The North Star:** Do you have a guiding vision that creates stability when the daily metrics fluctuate? If not, draft one today.

*****

The Gravity Leader

Our universe is full of stars, but only a few become gravitational legends — leaders whose pull shapes how others think, act, and dream.

In leadership research, this "pull" has a name: vision.

Scholars define visionary leadership as the ability to create a clear picture of the future that others can see, believe in, and emotionally connect to[x] . When a leader communicates a compelling future, people do not just follow instructions—they align their identities with the mission. They start to say, "This is our future."

That alignment is gravity.

Steve Jobs did not just sell computers. He framed technology as a tool for human creativity. Elon Musk did not just build cars; he framed Tesla as a step toward saving the planet. Research shows that when leaders articulate bold, meaningful visions, followers experience higher motivation and stronger commitment because they feel part of something larger than themselves.

But vision alone is not enough.

Another body of research focuses on meaningful work—the idea that people perform better when they believe their work matters. Meaningfulness in work refers to how much significance individuals feel their work holds in their lives[xi]. Leaders play a central role in shaping that meaning. They influence how people interpret setbacks, success, and daily effort.

In simple terms:

Leaders do not just assign tasks—they shape how people understand those tasks.

This helps explain philanthropists like Gandhi or Mother Teresa. Gandhi's Salt March was not about salt. It was about dignity. He transformed a simple act into a symbol of national self-respect. That reframing created moral gravity. People were not following a strategy; they were following meaning.

For example, when Atatürk introduced sweeping reforms in Turkey—from adopting the Latin alphabet to restructuring governance—he was not only modernizing administrative systems.

He was redefining Turkish identity from that of imperial subjects to that of citizens of a secular republic. That shift altered education, law, language, and culture. It was an institutional change driven by identity transformation.

This is constructive gravity.

But gravity has a shadow.

Psychological research distinguishes between healthy confidence and grandiose narcissism. The Narcissistic Admiration and Rivalry Concept explains that narcissism has two sides: admiration (seeking status through charm and vision) and rivalry (defending status through aggression and hostility)

In plain language:

- Admiration says, "Look at what I can build."
- Rivalry says, "No one is allowed to challenge me."

Studies show that leaders high in grandiose narcissism can initially appear bold and visionary, but when their self-image is threatened, they may become defensive and self-interested [xii]. Other research on CEOs shows that narcissistic leaders often "surface act"—projecting confidence while internally masking insecurity or isolation[xiii].

This explains why some leaders inspire breakthroughs while others create breakdowns.

Elizabeth Holmes is a clear example. Her vision—instant blood diagnostics from a single drop—was magnetic. Investors, employees, and media were drawn in. But when reality threatened the vision, transparency gave way to denial. Research shows that when narcissistic leadership becomes driven by self-interest rather than shared purpose, it predicts workplace deviance and destructive outcomes.

Vision without ethical grounding becomes dangerous gravity.

This brings us to ethical leadership.

Ethical leadership research defines ethical leaders as those who model fairness, encourage open communication, and define success not just by results, but by how those results are achieved[xiv]. Studies show that ethical leaders protect team confidence and cohesion even when performance is poor.

Why does this matter?

Because gravity is tested most during failure.

When teams lose, confidence erodes. Social bonds weaken. People begin blaming one another. But ethical leaders stabilize the orbit. They help teams interpret setbacks as opportunities for learning rather than as a collapse. They prevent rivalry from replacing trust.

This is why some bold leaders build enduring institutions — and others implode.

- Gravity Leadership is not charisma.
- It is not boldness alone.
- It is not domination.

It is the disciplined integration of:

- Vision (where we are going)
- Meaning (why it matters)
- Ethics (how we get there)
- Humility (when to adjust course)

From Gandhi's moral force to Musk's technological ambition, gravity takes different forms. But in every case, true gravitational leaders create alignment, not fear. They create shared purpose, not personal mythology.

Real gravity does not just attract attention.

It holds systems together.

Light Year Lessons	"As someone who's fascinated by data, I thought algorithms were the answer to every problem. But I've realized that its purpose—what brings people together and keeps us committed—that is true gravity. Data makes us smarter; purpose makes us unstoppable." —Alice

The New Frontier of Tech Leadership

The advent of AI has introduced a new dimension to this gravity leadership paradigm. This necessitates a shift from mere technical oversight to strategic leadership. Modern tech leaders must now possess the foresight to identify opportunities where AI can deliver significant business value, the acumen to guide teams through the complexities of AI development and deployment, and the adaptability to pivot strategies as AI technologies mature and evolve. This requires a more nuanced understanding not just of how technology works, but also of why it is being implemented and what strategic objectives it serves.

Gone are the days when a manager's primary responsibility was confined to ensuring systems ran smoothly, projects were delivered on time and within budget, or technical debt was managed. While these operational imperatives remain crucial, they are now intertwined with a much broader strategic mandate. The modern tech leader is no longer solely an overseer of technical execution but a strategic architect, a catalyst for innovation, and a navigator of complex, AI-infused ecosystems.

Bob's leadership focused on perfecting coding standards, optimizing build pipelines, and ensuring strict adherence to testing protocols—a master of process and precision.

But for Alice, the leadership landscape has shifted. Her daily orbit includes wrangling vast and messy data sets, selecting and fine-tuning complex models, and steering projects through the turbulence of algorithmic uncertainty. On top of that, Alice must grapple with ethical dilemmas that come with AI decision-making, where "right" is not always clear and the impact can stretch far beyond the server room.

In today's world, Alice's role demands a much broader skill set: she must blend technical brilliance with strategic foresight, exercise ethical judgment, and chart a course through uncharted technological frontiers. For leaders like Alice, success means going well beyond knowing how the tech works; it is about understanding why it is used and what direction it takes the organization.

In the chapters ahead, we will dive into the real-world challenges facing leaders like Bob and Alice, exploring both timeless obstacles and brand-new dilemmas born of the AI age. We will break down the specific strategies, mindsets, and practical plans that will empower both classic and next-generation managers to thrive in this rapidly changing world. Whether you relate more to Bob's purpose-driven approach or Alice's data-savvy instincts, you will discover actionable tools to help you—and your teams—navigate uncertainty, embrace innovation, and become the gravity that keeps everything moving forward.

*****

Summary

We have entered a new solar system of project management. The tools have changed, the speed has increased, and the risks are higher. But the laws of physics remain: teams need a center of gravity and a North Star to navigate the dark.

In the next chapter, we will look at **Mercury's Message**—how to craft that vision and communicate it with the speed required in an AI-native world.

2. Mercury's Message: Crafting Vision and Mission

Vision as Velocity

Bob always explains in this way: "Speed is just distance over time. Velocity is speed with a direction."

In ancient myths, Mercury was a Roman god, the messenger of the gods, known for his winged sandals and his impossible speed. In the modern AI era, Mercury represents the pace at which information moves. Algorithms trade stocks in microseconds; generative AI writes code in seconds; automated alerts flood our inboxes instantly.

For leaders, this speed is seductive. We are tempted to confuse motion with progress. We deploy, release, and pivot faster. But as Bob often reminds his team: "Moving at the speed of light doesn't help if you're flying into a black hole."

In the *Gravity of Command*, your Vision and Mission are not just corporate posters on a wall. They are the navigational vectors that turn raw speed into true velocity. Without them, AI helps you get lost faster.

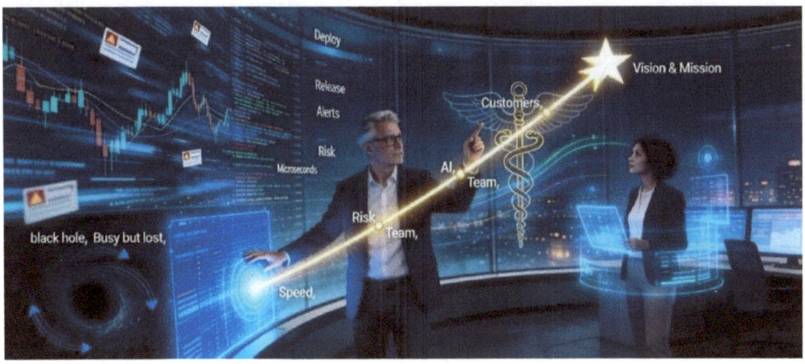

The Blackout: A Crisis of Communication

The difference between speed and direction became painfully clear to Alice and Bob during the "Great API Blackout" of last winter.

It started as a minor latency flag on a Tuesday morning—a “small detail” buried in a dashboard. By noon, a cascade of automated retry logic from an AI-driven client had overwhelmed the legacy authentication servers. The system did not just slow down; it crashed hard.

Alice’s team immediately went into hyperdrive. They were deploying hotfixes, rolling back containers, and spinning up new instances. They were moving incredibly fast but flailing.

Bob walked into the war room. The energy was frantic. Engineers were shouting across the room, pasting error logs into chat channels, and blaming the cloud provider.

“Stop,” Bob said. He did not shout, but his voice carried the weight of gravity. The room quieted.

“What is the mission right now?” Bob asked.

“To get the servers back up!” a junior engineer yelled.

“No,” Bob corrected. “The mission is to protect the customer’s data integrity. If we bring the servers up before we patch the leak, we corrupt the database. We are not racing the clock; we are protecting the asset.”

That single sentence realigned the room. The frantic speed turned into calculated velocity. The goal shifted from “fastest restart” to “safest recovery.” They lost an extra hour of uptime, but they saved the client’s data.

The Titanic Lesson: When Small Details Sink Big Ships

Later that evening, after the war room had emptied, Alice sat quietly.

“We were so focused on speed,” she said. “How did we miss something that small?”

Bob leaned against the table.

"Let me give you another picture," he said.

"You know the Titanic?"

Alice nodded.

"In its time, it was the most advanced ship ever built. Powerful engines. Reinforced compartments. Modern communication systems. People called it unsinkable."

He paused.

"It didn't fail because the engines stopped working."

"It failed because warning signals didn't translate into action."

Alice tilted her head.

"Imagine this," Bob continued. "The captain tells a crew member:

'Check the water temperature every ten minutes. If it drops suddenly, report immediately.'"

"That sounds minor," Alice said.

"Exactly," Bob replied. "Small. Repetitive. Boring."

"It's freezing outside. The crew member checks once. It's cold. Of course, it's cold. Next time, maybe he will delay it. Maybe he assumes nothing changed. Maybe he writes what he expects instead of what he measured."

"And then?" Alice asked quietly.

"And then the ship meets ice."

Silence.

Alice frowned. "So the crew member failed."

Bob shook his head.

"Maybe. But the deeper question is this: Did he understand why the task mattered?"

He let that sit.

"Leadership is not giving instructions. It's transferring importance."

He walked toward the screen, which still displayed the early latency spike from that morning.

"That was our water temperature."

Alice stared at it differently now.

"It looked insignificant," she said.

"Yes," Bob replied. "Because no one connected it to consequence."

"In complex systems, small signals rarely look dramatic. They look routine."

He continued:

"The Titanic wasn't sunk by a lack of horsepower. It was sunk by misaligned attention."

"In our case, the engines were fine. The cloud was fine. The technology worked."

"What failed was signal interpretation."

- **The AI Blind Spot:** We trust our automated dashboards so much that we stop "checking the water temperature" ourselves.
- **The Communication Gap:** When a junior engineer spots a "minor" anomaly in the training data, do they feel safe enough to stop the launch? Or do they stay silent to keep up the speed?

Mercury's Lesson is clear: You must build a culture where the smallest signal can travel to the captain instantly, without friction. Speed of communication saves ships; silence sinks them.

*****

- **Vision is Vector:** Speed without a defined mission is just chaos. Define the "North Star" before you hit the accelerator.
- **The Speak-Up Culture:** The fastest communication channel isn't Slack; it's psychological safety. If your team fears being the "bearer of bad news," your radar is broken.
- Small Data Matters: AI processes Big Data, but humans must catch the "Small Data"—the anomalies, the gut checks, and the temperature readings.

*****

Strategies for Handling Challenges for New Age Leaders

"Vision is the art of seeing what is invisible to others." – Jonathan Swift.

The cost of poor execution is staggering. In 2022, tech failures cost U.S. companies $2.41 trillion. Now fast-forward to 2026: [2]Gartner reports that 60% of AI projects are abandoned mid-implementation due to data quality issues. This is not just a statistic—it is your Titanic moment.

The MIT GenAI study quantified it sharply: 95% of enterprise generative AI pilots fail to achieve ROI within six months. Why? Because leaders focused on the AI (the technology) instead of the North Star (the outcome). A common failure pattern emerges:

Top Five Reasons GenAI Pilots Fail, as per Gartner 2025 studies, are:

1. No clear ROI definition (37%) — 'improve productivity' is not a target
2. Data quality issues (28%) — training on garbage yields garbage
3. Integration complexity (21%) — the pilot lives in isolation

[2] *Gartner AI Failure Analysis (2025); Gartner Trough of Disillusionment Forecast (Jan 2026)*

4. Change management gaps (14%) — your team hasn't bought the North Star
5. Leadership misalignment (McKinsey finding) — departments optimize locally, not globally

Small details. Massive consequences. Your role: catch them before they snowball.

You might be surprised to know that annual maintenance costs for software often exceed the initial development expenses. Approximately 90% of companies' IT budgets are spent on software maintenance and support, including defect fixes, software updates, and ensuring compatibility with other systems.[3]

This indicates that, in today's landscape, speed is not just an advantage; it is a necessity. Mercury, the fleet-footed messenger of the cosmos, reminds us that successful leaders do not just move quickly; they deliver the right messages to the right people at the right moment. For modern leaders like Alice and Bob, forging a clear vision and mission is the new essential skill for navigating constant change, inspiring teams, and aligning everyone around a common destination.

Resilience Through Purpose

Gravity Leadership shines most during tough times, providing teams with a sense of stability and direction. When challenges hit, purpose-driven leaders help their teams stay grounded and resilient, creating a shared purpose that anchors everyone and builds confidence to face uncertainty together.

Bob recalls one particularly challenging project. With a key deadline fast approaching, the client made an urgent change request that would require the team to work through the weekend to keep things on track. Bob understood how critical the delivery was for the client's business and immediately opened a conversation, seeking ways to meet the goal without overtaxing the team.

---

[3] *MIT GenAI Divide Study (2025);*

Bob spoke with both the client and the team, acknowledging the importance of the deadline while also expressing his respect for the team's Diwali celebrations. He offered to explore creative options—from rebalancing tasks to seeking temporary support elsewhere—trying to build a plan that honored the needs of both sides.

When no practical workaround could be found, Bob made his priorities clear: he let the client know that the team's well-being and cultural values mattered. He explained his decision respectfully, emphasizing that a motivated and well-rested team would always serve the client's interests better in the long run. The client, though disappointed and initially frustrated, appreciated Bob's honesty and commitment.

To Bob's surprise, several team members, inspired by this trust and openness, stepped up voluntarily. They chose to help even though they knew their leader would support any decision they made. This spirit of shared purpose did not just deliver the project on time; it strengthened the team's unity and established lasting trust with the client, who saw firsthand how much Bob valued partnership, transparency, and results.

The principle holds just as true in today's high-stakes technology world. Let's look at some well-known incidents and software release failures:

- **Major Outages:** Infamous events, like the 2016 outage at Delta Airlines or broken updates at Microsoft, have been traced back to minor configuration errors, missed test cases, or overlooked security warnings.
- **AI Model Bias:** Modern AI projects can go off course if teams overlook "small" steps, such as properly balancing the training data or validating edge cases. These oversights can snowball into biased outcomes, lost trust, and costly rework.
- **Cloud Migrations:** Many cloud failures are traced to skipped process checklists or misconfigured permissions. A single unchecked box can disrupt entire businesses.

No matter the domain, the lesson is clear: in complex, fast-moving environments, small details can be the difference between smooth sailing and disaster.

 Mars Rovers	Checklist: Avoiding Titanic Moments for Leaders and Project Teams

Crafting Your Vision in the AI Era

In a world where AI can generate a mission statement in seconds, your job is not just to write it. Your job is to embody it.

1. **Be Brief (Mercury's Rule):** If your vision takes three paragraphs to explain, it is too slow for the AI age. It should be a single vector. For example: "We protect the data, no matter the cost."
2. **Be Human:** AI communicates data; leaders communicate meaning. Use stories, analogies, and face-to-face time to make the vision stick.
3. **Be Consistent:** The vision cannot change with every gust of wind. It must be the gravity that holds the team when the technology fails.

*****

The Communication Velocity Check

- [] The **"Pre-Mortem":** In your next planning session, ask: "If this project fails, what 'small detail' did we likely ignore?"
- [] **The Hotline:** Establish a "Red Flag" channel (or keyword) that anyone, regardless of rank, can use to pause a project without penalty.
- [] **The Elevator Pitch:** Can your newest hire explain the business value of their current task in 30 seconds? If not, the vision has not landed.

*****

Mission Control: Generating Momentum

- Regularly review and communicate the "critical details" that can make or break your project.
- Foster a culture where everyone feels safe to report minor issues or risks, no matter how trivial they seem.
- Ensure checklists exist not only for launch or go-live, but also for every phase, including routine maintenance and disaster recovery.
- Empower and train team members at all levels to understand why their tasks matter for the bigger picture.
- Celebrate those who spot and fix minor issues before they become significant problems.
- Avoid overconfidence: review lessons learned from past misses, both within your team and from industry headlines.
- Appoint "detail champions" for key risk areas, especially in complex or multi-team efforts.
- Conduct "pre-mortems": ask the team what small details, if missed, would sink the mission.
- The smallest action, when done with care and purpose, helps steer the entire mission safely forward.

Commitment vs. Compliance

What sets truly effective leaders apart is their ability to inspire genuine commitment, rather than settling for mere compliance. A compliance-focused leader assigns tasks without explanation, leaving team members disengaged and unsure why their work matters. In contrast, a purpose-driven leader connects each task to the bigger mission, sparking pride and a sense of ownership in their team.

Consider an airline maintenance technician inspecting rivets on an aircraft. To an outsider, this might look like tedious, routine work on a long checklist. But history shows that overlooking even a single small flaw can have catastrophic consequences. When leaders stress the life-or-death importance of these checks, they transform an ordinary job into an essential one. This shift in perspective encourages greater vigilance and a culture of excellence.

Fostering a Culture of Accountability

The Titanic's tragedy underscores the need for a workplace where accountability rests with everyone, not just those at the top. Leaders foster this by:

- **Emphasizing the Why:** Regularly explain how each task supports the overall mission. Make sure every team member knows their work makes a difference.
- **Modeling Attention to Detail:** Show the importance of small actions by treating details with care yourself. When leaders set this example, the team will follow.
- **Recognizing Contributions:** Acknowledge the behind-the-scenes heroes. Highlighting overlooked roles reinforces that every task is essential.
- **Encouraging Proactive Problem-Solving:** Urge team members to speak up when they spot potential problems, no matter how minor.

By leading with purpose, detail, and respect, leaders create environments where commitment and shared responsibility come naturally and are sustained over time.

*****

Summary

Mercury teaches us that communication is the nervous system of any project. In the age of AI, where information moves instantly, the leader's role is to ensure that information has meaning.

You must filter the noise, amplify the critical signals (even the small ones), and ensure that every burst of speed is aligned with your North Star. In the next chapter, **Venusian Balance**, we will explore how to harmonize this need for control with the chaotic beauty of AI-driven innovation.

3. Venusian Balance: Command and Control in Harmony

Venusian Balance: The Tug of War

"The machine says we can deliver in six weeks," Alice said, pointing to the Gantt chart generated by the new AI project management tool. "It analyzed our last fifty sprints, factored in holidays, and optimized the critical path. It's perfect."

Bob crossed his arms, staring at the screen. "It's efficient, Alice. But it's not perfect. Look at the resource allocation for the backend team. It has assigned the heaviest integration tasks to the Offshore Bengaluru unit during the week of Diwali."

Alice frowned, tapping the screen. "I checked the system settings. It's not flagged as a 'Non-Working' block in the global corporate calendar, so the AI treated it as available capacity."

"That's exactly the problem," Bob replied. "The algorithm sees 'available hours.' It does not see a cultural event that is as significant to them as Christmas is to us. If we run this schedule, we aren't just missing a deadline; we are signaling that we don't respect their culture. We lose the team before we even start."

Alice paused. She had not caught that. The algorithm saw "calendar data." Bob saw "human context."

This dynamic represents the central challenge of the Venusian Balance. In the solar system, Venus is a planet of extremes: crushing pressure and intense heat. In project management, the pressure comes from the opposing forces of **Control** (compliance, metrics, structure) and **Command** (influence, vision, empathy).

In the age of AI, the temptation is to let the machine take "Control" of everything. But as Bob taught Alice that day: "AI gives us the science of the schedule. Leadership provides the art of the adjustment."

Figure 1.2: The balance between managing processes (Control) and managing Command) for organizational success.

Command vs. Control: The AI Distinction

To master this balance, we must first distinguish between two often-confused terms.

- **Control** is about enforcement. It is binary. Did the task get done? Is the budget green? Is the server online? AI excels here. It can monitor compliance, enforce code standards, and track variances faster than any human.
- **Command** is about empowerment. It is nuance. It is guiding a team through ambiguity, making ethical judgment calls, and inspiring ownership. AI struggles here.

The Venusian Rule: Give the Control to the AI. Keep the Command for yourself.

When managers try to "control" people through micromanagement or "command" data by ignoring facts, projects fail. The goal is harmony: use AI to tighten the nuts and bolts so you can steer the ship.

The New Planning: From Guesswork to Prediction

Historically, project scoping was a dark art. We held meetings, made guesses, added a 20% "optimism buffer," and hoped for the best.

Today, AI transforms planning from a reactive exercise to a predictive science.

1. **The End of "Optimism Bias":** Humans are terrible estimators. We forget past pain. AI remembers everything. By ingesting historical data from your ticketing systems, AI tools can predict task duration based on actual past performance, rather than on what a developer hopes will happen.
   - **Alice's Win:** Instead of asking, "How long will this take?", Alice asks the AI, "How long did this take us last time?" The AI might say, "Similar tasks averaged 14 days, not the 5 days estimated by the engineer."
2. **Scoping with NLP:** Requirements gathering is often messy. Natural Language Processing (NLP) tools can now scan thousands of customer support tickets, emails, and user feedback logs to identify the "real" requirements versus the "assumed" ones.
   - **The Result:** You do not just build what stakeholders say they want; you build what the data proves they need.

*****

AI can help you enforce quality without being the "bad guy."

- **The Old Way:** The manager argues with the developer about whether the code is ready.
- **The AI Way:** The AI analyzes the code against the "Definition of Done" benchmarks (test coverage, security scans, and documentation). If it does not pass, the ticket does not close. The standard is objective, not personal.

*****

Mastering AI Integration in Project Management

The advent of AI is not merely a technological upgrade; it represents a profound paradigm shift in how we approach every facet of project management, beginning with the crucial initial stages of planning and scoping. Traditionally, these phases have relied heavily on human expertise, experience, and, often, intuition. While these human elements remain invaluable, AI introduces an unprecedented layer of data-driven precision, predictive power, and efficiency.

Imagine moving from meticulously compiling project requirements through manual interviews and surveys to a system that can intelligently analyze vast repositories of user feedback, market research, and even competitive product documentation to surface potential needs and opportunities. This is the promise of AI-enhanced project planning and scoping.

Intelligent Execution: The Resource Puzzle

One of the most complex parts of Bob's job was resource allocation—playing Tetris with people's schedules. Resource allocation, a cornerstone of effective project planning, traditionally involves a complex balancing act that considers individual skill sets, availability, project dependencies, and budget constraints.

AI changes this by moving from "availability-based" assignment to "skill-based" optimization. It optimizes this process by analyzing an organization's entire resource pool, understanding the specific demands of each project task, and factoring in historical performance data.

- **Example:** If a project requires a highly specialized software engineer with experience in a particular legacy system, AI can quickly identify individuals with that expertise and availability within the required timeframe, even suggesting alternative resources.
- **Beyond Lookups:** Modern tools can analyze a team member's commit history and recent training to see who has the best skill for a task, not just who has the time. If a task requires a specific legacy JavaScript framework, the AI can identify that "David" worked on a similar module two years ago, even if his current title does not reflect it.

The Human Guardrail: This is where Bob's intervention regarding Diwali was critical. The AI optimizes for efficiency (throughput). The leader must optimize for sustainability (culture). If you blindly follow the AI's resource plan, you risk alienating global teams by ignoring the cultural events that define their lives. You must command the system to respect the human calendar.

Analysis at Scale

At its core, AI's contribution to planning and scoping is rooted in its remarkable ability to process and interpret complex datasets at speeds and scales far beyond human capacity. Consider the intricate process of requirements gathering.

Instead of Bob painstakingly synthesizing notes from countless meetings, Alice can ingest and analyze textual and verbal data from stakeholder interviews, customer support logs, and internal documentation. Natural Language Processing (NLP) algorithms can identify recurring themes, potential ambiguities, and even stakeholder sentiment, flagging areas that require further clarification.

This not only accelerates the identification of core requirements but also helps uncover nuanced needs that might be overlooked in a manual process. Furthermore, AI can cross-reference these identified requirements against existing product roadmaps, market trends, and technological feasibility studies. This provides an early assessment of viability and impact, helping to define a project scope that is both aligned with immediate needs and strategically positioned for future success.

Risk Management: The Pre-Mortem

Bob's traditional risk management was a "Risk Register"—a spreadsheet where he would list potential issues and then forget about them. Alice, however, uses Predictive Risk Intelligence.

Imagine a system that continuously scans your project—monitoring supply chain news, developer commit rates, and budget burn rates—to flag risks before they materialize.

- **The Scenario:** An AI tool notices that a third-party API vendor has delayed their last three releases. It flags your project's dependency on that vendor as "High Risk" weeks before your integration phase begins.
- **The Action:** You do not wait for the delay. You build a contingency stub or switch vendors now. You move from firefighting to fire prevention.

The parallels to advanced data analytics in technical troubleshooting are striking. Just as AI can sift through mountains of logs and error reports to pinpoint the root cause of a system failure, it can now analyze historical project data to anticipate potential pitfalls in new ventures.

This includes identifying patterns associated with scope creep. By examining past projects that experienced uncontrolled expansion, AI can highlight specific triggers, communication breakdowns, or stakeholder behaviors that are statistically correlated with scope creep. Armed with these insights, project managers can proactively implement preventative measures, such as more rigorous change control processes, clearer definitions of "out of scope" items, or targeted stakeholder engagement strategies. This predictive capability transforms risk management from a reactive to a preemptive exercise, allowing teams to build resilience into the project plan from the very beginning.

Integrating AI into Established Frameworks

Integrating these AI-driven insights into established project management frameworks, such as Agile or Waterfall, is crucial for maximizing their impact.

- **In an Agile Environment:** AI can continuously refine backlog prioritization. By analyzing user feedback, market performance of existing features, and development velocity, AI can recommend which user stories are most critical, which require further refinement, or which might be deprioritized due to evolving market conditions. This allows Agile teams to maintain a more dynamic and responsive backlog. For instance, if an AI analyzes a sudden surge in customer inquiries about a specific feature, it can immediately flag this for the product owner to consider for an upcoming sprint.

- **Within a Waterfall Methodology, AI's influence is strongest** in the initial planning and design phases. The detailed project plan can be populated with AI-generated estimates for task durations, resource requirements, and potential risks. AI can also assist in creating more robust Work Breakdown Structures (WBS) by analyzing similar past projects and suggesting optimal task decompositions, ensuring that no critical work packages are overlooked.

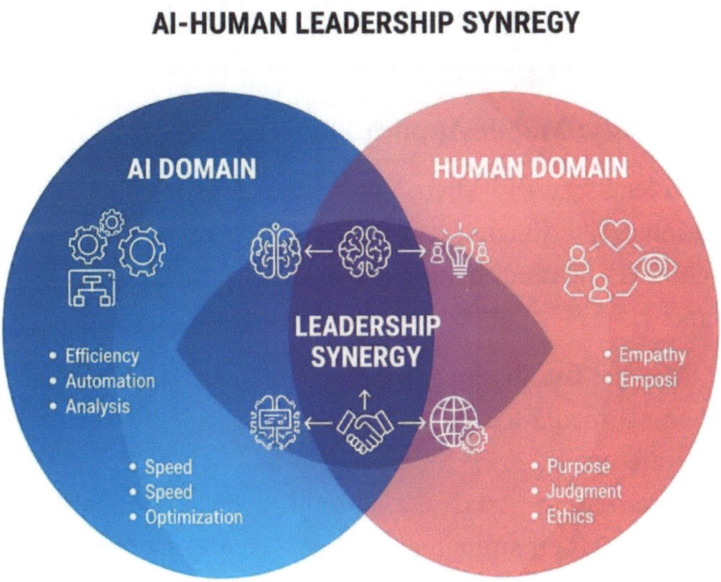

Figure 1.3: The collaborative overlap of AI and Human intelligence for synolesstic lendership

The Augmented Project Manager

The concept of "AI-enhanced" does not imply replacing the project manager; rather, it augments their capabilities. The human element remains indispensable for strategic decision-making, ethical considerations, and fostering team cohesion. However, by leveraging AI, project managers can transition from spending excessive time on data aggregation to focusing on higher-value activities such as strategic alignment and complex problem-solving.

Furthermore, AI can facilitate a more dynamic and iterative approach to scoping. While Waterfall traditionally emphasizes upfront, fixed scoping, AI can introduce elements of continuous refinement. As initial development progresses, AI can analyze early user testing results and competitor movements to suggest minor, data-backed adjustments to the scope that might enhance the project's overall success without derailing the core plan.

Case Study: Mobile Application Development

Consider a scenario where a project aims to develop a new mobile application. Traditionally, scoping would involve extensive market research and detailed feature definition. With AI, the process is accelerated:

1. **Market Intelligence:** An AI could ingest data from app stores, social media trends, and tech news to identify emerging user preferences.
2. **Competitive Analysis:** It could analyze competitor apps to identify their strengths, weaknesses, and user reception.
3. **Predictive Modeling:** The AI could generate a preliminary feature set, predict user adoption rates, and suggest optimal pricing models.

The project manager and their team then review these AI-generated insights, using their experience to validate and finalize the scope. This collaborative approach ensures that the project is both innovative and grounded in practical realities.

The benefits extend to cost estimation as well. By analyzing historical data, including actual costs incurred for similar tasks, AI can generate more accurate cost projections. It can identify cost drivers, predict potential budget overruns based on early indicators, and suggest cost-saving alternatives. This enhanced accuracy allows for more realistic budgeting, reducing the likelihood of unexpected financial challenges later in the project lifecycle.

Moreover, AI can play a vital role in defining the "definition of done" for various project deliverables. By analyzing past projects and industry best practices, AI can help establish objective, measurable criteria for completion, ensuring that quality standards are consistently met. This is particularly useful for complex software projects where "done" can be a subjective concept. AI can help quantify metrics such as code coverage, performance benchmarks, and security vulnerabilities, providing clear, data-backed indicators of completion.

Implementing AI in project planning and scoping poses its own challenges. It requires access to high-quality, well-structured historical data. Organizations that have poor data governance or incomplete project records will find it challenging to leverage AI effectively. Furthermore, there is a learning curve associated with adopting these new tools and integrating them into existing workflows. Project managers and team members need to be trained on how to interpret AI outputs, understand their limitations, and use them to inform their decisions.

However, the potential rewards are substantial. AI-enhanced planning and scoping lead to more accurate estimates, reduced risk of scope creep, optimized resource allocation, and ultimately, a stronger foundation for project success. It empowers project managers to move beyond the reactive and embrace a proactive, data-informed approach, transforming the often-arduous initial phases of a project into a more efficient, precise, and strategic endeavor. This early success in planning sets the stage for smoother execution and a higher probability of achieving the desired project outcomes, all while freeing up valuable human capital for more critical strategic thinking and problem-solving. The integration of AI into these foundational stages is not just an evolutionary step; it is a revolutionary leap forward in the discipline of project management.

To elaborate further on the predictive capabilities of AI in scoping, consider the concept of "epics" and "features" within product development, particularly in Agile contexts. Traditionally, defining these can be an iterative process involving numerous workshops and stakeholder consensus-building. AI can accelerate this by analyzing user journey maps, customer support tickets, and competitor feature sets to propose potential epics and their constituent features. It can then go a step further by predicting the likely effort required for each feature, the potential user impact, and the strategic alignment with broader business objectives.

For example, if a project aims to build an e-commerce platform, AI could analyze the typical user flow from discovery to purchase, identify friction points in existing platforms, and suggest features to address these, along with estimated development complexity and potential ROI. This allows product owners and project managers to make more informed decisions about what to prioritize and how to break down large initiatives into manageable components, ensuring that the project scope is both ambitious and achievable.

The ability of AI to analyze vast amounts of unstructured data is also a game-changer for requirements gathering. Imagine a scenario where a company is developing a new enterprise software solution. Instead of relying solely on interviews with a limited number of key stakeholders, an AI can be tasked with analyzing all internal communication channels (with appropriate privacy controls), relevant industry whitepapers, competitor product documentation, and even public forums where potential users discuss their pain points.

The AI's NLP capabilities can then identify recurring themes, unmet needs, and desired functionalities. It can flag inconsistencies in stakeholder requests or highlight potential conflicts in requirements.

This comprehensive analysis provides a richer, more nuanced understanding of user needs and business requirements, leading to a more accurate and complete project scope definition. The project manager then uses their judgment to synthesize this AI-generated intelligence, validating it with stakeholders and ensuring that the final scope aligns with strategic goals. This augmentation of human analysis with AI's broad scanning capability provides a more robust and well-informed starting point for any project.

The realm of risk assessment is particularly ripe for AI-driven transformation in project planning. Beyond identifying risks based on historical data, AI can engage in more sophisticated forms of risk prediction. For instance, by analyzing real-time news feeds, economic indicators, and geopolitical developments, AI can flag emerging risks that might impact a project's supply chain, regulatory environment, or market demand. This is especially pertinent for projects with long lead times or those operating in volatile industries.

An AI might detect an impending trade dispute that could affect the cost of imported components, or an upcoming legislative change that could impact the compliance requirements of a software product. These real-time alerts allow project managers to proactively adjust plans, explore alternative sourcing, or modify project specifications before these risks materialize into significant issues. This level of continuous risk monitoring, powered by AI, provides an unprecedented level of foresight, allowing for more agile and resilient project planning.

Furthermore, AI can help quantify risks. While traditional risk registers often use qualitative scales (e.g., low, medium, high) for probability and impact, AI can leverage historical data to assign more precise probabilities and financial implications to identified risks. By analyzing thousands of similar past projects, AI can provide data-backed estimates for the likelihood of a particular risk occurring and the potential financial or schedule impact if it does. This granular level of risk quantification enables project managers to make more informed decisions about risk mitigation strategies, contingency planning, and resource allocation for risk management activities. It moves risk management from an art to a more precise science, grounded in empirical data.

In terms of resource allocation, AI can optimize not only who gets assigned to a task but also when and how. For example, in a complex software development project with numerous interdependencies, AI can analyze task durations, resource availability, and skill requirements to create an optimized critical path, identifying potential bottlenecks before they emerge. It can also suggest the most efficient task sequencing to minimize idle time for specialized resources. Beyond simply assigning individuals, AI can recommend the optimal team composition for specific tasks, considering factors such as individual strengths, past collaboration success rates, and even personality compatibility scores, provided such data is ethically collected and used. This holistic approach to resource management ensures that the right people are in the right place at the right time, maximizing productivity and minimizing project delays.

The integration of AI into project planning and scoping also facilitates a more objective and data-driven approach to defining project success metrics. While traditional metrics might focus on on-time and on-budget delivery, AI can help identify and track a broader set of performance indicators. For example, in a customer-facing project, AI can analyze post-launch customer feedback, user engagement data, and market share to provide a more comprehensive assessment of project success beyond the initial launch. This allows for a continuous feedback loop, where insights from the project's performance inform future planning and scoping, creating a cycle of constant improvement.

The iterative nature of AI-driven planning is particularly beneficial in today's rapidly evolving technological landscape. Instead of a rigid, upfront scope that may quickly become obsolete, AI allows for a more dynamic and adaptable planning process. As new information emerges—perhaps a breakthrough in a related technology, a shift in competitor strategy, or a change in customer preferences—AI can quickly re-evaluate the project plan and scope. It can suggest adjustments to features, timelines, or resource allocations to ensure the project remains aligned with current realities and future opportunities. This adaptability is crucial for organizations aiming to stay agile and competitive in sectors characterized by rapid innovation and disruption.

Consider the construction industry, an area often perceived as traditional but increasingly adopting advanced technologies. AI can revolutionize the scoping and planning of large construction projects. By analyzing geographical data, soil conditions, historical construction data for similar sites, and even weather patterns, AI can predict potential construction challenges, optimize site layouts, and generate highly accurate cost and timeline estimates. It can identify optimal material sourcing strategies, expect potential delays due to environmental factors, and suggest the most efficient construction sequences. This level of detailed, data-driven planning, powered by AI, significantly reduces the inherent risks in large-scale construction projects, leading to more predictable outcomes and improved efficiency.

Furthermore, the ethical considerations of AI in planning are paramount. When using AI for risk assessment or resource allocation, it is crucial to ensure algorithms are free of bias. For example, if historical project data reflects past discriminatory hiring practices, an AI trained on this data might perpetuate those biases in resource allocation recommendations. Project managers must be vigilant in scrutinizing AI outputs for fairness and equity, ensuring that decisions are made ethically and responsibly. This involves understanding the data sources and the algorithms used, while implementing checks and balances to mitigate potential biases. Human oversight remains critical to ensure that AI is used as a tool to enhance decision-making, not to dictate it in an unexamined manner.

Ultimately, integrating AI into project planning and scoping is about enhancing human capabilities through intelligent automation. It is about moving beyond guesswork and intuition to a more precise, predictive, and data-informed approach. By leveraging AI's power to analyze vast datasets, identify patterns, and make predictions, project managers can develop more robust, realistic, and adaptable project plans from the outset. This not only increases the likelihood of project success but also frees up valuable human cognitive resources to focus on strategic thinking, innovation, and stakeholder engagement, thereby elevating the practice of project management to new heights.

The initial phase of a project, often fraught with uncertainty, can be transformed into a period of clarity and confidence, setting a strong foundation for everything that follows. This proactive, intelligent approach to planning ensures that projects are not just initiated but are set on a trajectory for meaningful and impactful outcomes.

The evolution of project management is intrinsically linked to the tools and methodologies we employ. As we move beyond the foundational stages of planning and scoping, the immediate challenge becomes the efficient execution and diligent oversight of the myriad tasks that constitute a project. This is where AI-driven task management and automation emerge as transformative forces, shifting project execution from an often labor-intensive, reactive process to one characterized by intelligent foresight, dynamic adaptation, and streamlined operations. The core promise here is not merely to manage tasks but to optimize the flow of work, ensuring that every team member is engaged in activities that maximize their contribution and that the project progresses with unprecedented efficiency.

At the heart of AI-driven task management lies its remarkable capacity for intelligent assignment and resource optimization. Traditional task assignment often relies on a project manager's knowledge of team members' availability and skill sets, which can be prone to manual oversight, outdated information, or simply the sheer volume of tasks overwhelming even the most organized individual.

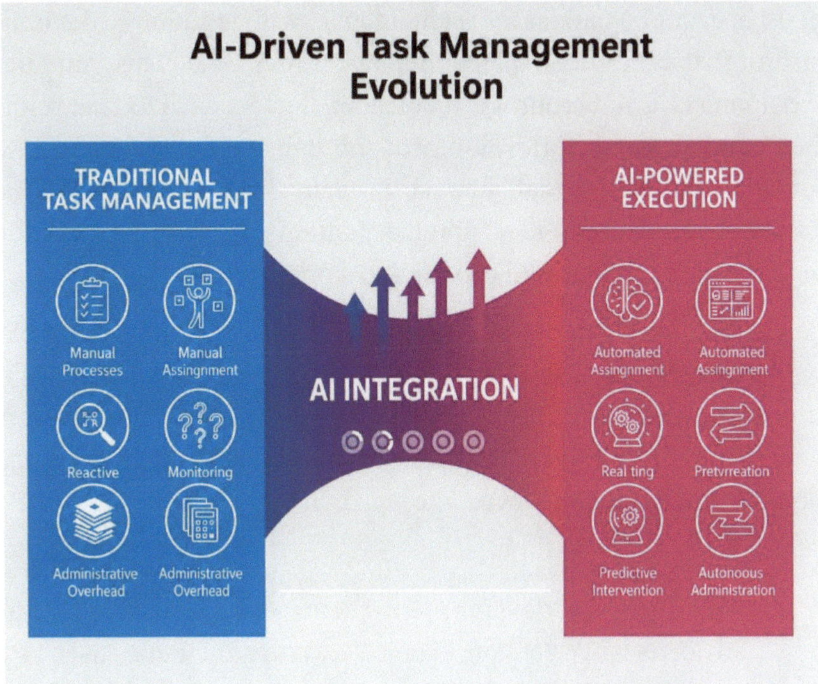

AI, however, can ingest and continuously update data on individual team members' capacities, skill proficiencies (even at granular levels of expertise), current workloads, and preferred working styles. By analyzing the requirements of a specific task—its complexity, required skills, estimated duration, and dependencies—an AI system can recommend or even automatically assign that task to the most suitable individual. This goes beyond simple keyword matching; sophisticated AI can infer skill relevance from past project contributions, completed training modules, and contributions to open-source projects.

For instance, if a project requires a developer with specific expertise in a particular JavaScript framework for a front-end component, the AI can identify the team member who not only possesses that skill but has also recently completed a related task, ensuring technical competence and recent experience. This dynamic reassignment capability becomes particularly powerful when unforeseen circumstances arise. If a team member falls ill or a critical deadline shifts, the AI can instantly reassess the project's task landscape and reallocate assignments with minimal disruption, ensuring that critical paths remain on track. This intelligent reassignment is not arbitrary: it considers factors such as task priority, the potential for skill development through a challenging task with guidance, and the overall impact on team dynamics. The goal is to maintain momentum and prevent bottlenecks before they form, ensuring a continuous and productive workflow.

Beyond assignment, AI significantly enhances progress tracking by moving beyond simple check-ins and manual updates. While project managers have traditionally relied on team members to report their progress, AI can facilitate more objective and continuous monitoring. This is achieved through integration with the tools the team already uses.

- **For software development teams:** AI can analyze commit histories in version control systems, track user story completion in task boards like Jira or Asana, and monitor the frequency and success rate of automated tests.
- **For design teams, AI might analyze progress** on collaborative design platforms or the completion of specific design milestones.

This continuous data stream enables the AI to provide real-time updates on task status, identify tasks that are falling behind schedule, and flag potential impediments with remarkable accuracy. For example, if a developer's progress on a critical feature stalls, the AI can detect this by observing a lack of code commits, stagnation in task board updates, or recurring failures in automated builds. It can then proactively alert the project manager and offer potential solutions, such as suggesting additional resources or identifying dependencies causing the delay. This real-time visibility transforms progress tracking from a retrospective activity into a predictive and preventative one, allowing for early intervention rather than last-minute crisis management.

The automation of repetitive and time-consuming tasks is another cornerstone of AI's impact on project execution. Project managers and team members often find themselves bogged down by administrative overhead: generating reports, scheduling meetings, sending reminders, and collating data. AI can automate many of these functions, freeing up valuable human capital for more strategic and creative endeavors.

For instance, the generation of daily, weekly, or monthly status reports can be fully automated. An AI can pull data from various project management tools, synthesize key metrics (tasks completed, milestones achieved, risks identified, budget consumed), and present this information in a customizable report format. This not only saves countless hours but also ensures consistent, accurate reporting. Similarly, AI-powered tools can manage meeting scheduling by analyzing participants' calendars to suggest optimal meeting times, sending invitations, and creating agendas based on ongoing project discussions or upcoming deadlines. Automated reminders for upcoming deadlines, pending approvals, or required actions can be sent intelligently, reducing cognitive load on team members and minimizing the risk of tasks slipping through the cracks. This automation of the mundane allows individuals to focus their energy on problem-solving, innovation, and collaborative efforts that truly drive project value.

Furthermore, AI can play a crucial role in identifying and resolving bottlenecks within project workflows. By analyzing the flow of tasks, dependencies, and resource allocation, AI can pinpoint specific points in the process where work tends to accumulate or slow down. This could be a particular team member who is consistently overloaded, a specific type of task that frequently encounters delays, or a dependency on an external team or vendor that is often late.

Once identified, the AI can not only flag these bottlenecks for the project manager but also suggest potential solutions. For example, if a bottleneck is identified due to a shortage of a specific skill, the AI might suggest reallocating tasks, bringing in external support, or prioritizing training for existing team members. If the bottleneck is specific to a workflow, the AI could propose optimizing the task sequence, streamlining approval processes, or introducing parallel processing where possible. This continuous analysis and optimization of workflows ensures that the project remains fluid and efficient, adapting to changing demands and preventing stagnation.

The concept of workflow automation extends to more complex scenarios, such as the management of approvals and reviews. In many projects, especially those involving regulatory compliance or quality assurance, a series of approvals is required before a task or deliverable can be considered complete. AI can streamline this process by intelligently routing documents to the appropriate approvers based on predefined criteria, tracking approval statuses, and automatically escalating items approaching their due dates or that have been pending for too long. This ensures that the review and approval process does not drag on project progress, keeping the project moving efficiently.

For example, in a pharmaceutical R&D project, AI could manage the review of experimental protocols, ensuring each is reviewed by the relevant subject matter experts, statisticians, and regulatory affairs personnel in the correct sequence, and flagging any deviations or potential issues early in the process.

Another significant application of AI in task management is its ability to predict potential future challenges and proactively mitigate them. By analyzing historical project data, including past successes, failures, and common issues encountered, AI can identify patterns that may indicate future problems. For instance, if past projects with similar complexity and team compositions have consistently experienced delays in a particular phase, the AI can alert the current project manager to this risk and suggest preemptive measures. This might involve allocating additional resources to that phase, conducting more frequent check-ins, or preparing contingency plans. This predictive capability transforms risk management from a reactive exercise of responding to problems as they arise into a proactive strategy of anticipating and preventing them. This foresight is invaluable in complex, long-term projects where even minor delays can have significant ripple effects.

The integration of AI into task management also fosters greater team satisfaction and engagement. When mundane, repetitive tasks are automated, team members are freed to focus on more engaging, intellectually stimulating work. This can lead to increased job satisfaction, higher levels of creativity, and a greater sense of accomplishment. Furthermore, intelligent task assignment, which ensures individuals work on tasks aligned with their skills and interests, can lead to better performance and a more positive work environment. When team members feel that their contributions are valued and that they are working on challenging yet achievable tasks, their overall morale and productivity tend to improve. The AI-driven system acts as an intelligent assistant, supporting the team rather than simply dictating assignments, thereby empowering individuals and fostering a more collaborative, productive environment.

Consider the implications for distributed and remote teams. In such environments, maintaining visibility and ensuring seamless collaboration can be particularly challenging. AI-driven task management tools can bridge these gaps by providing a centralized, transparent platform for task assignment, progress tracking, and communication. Real-time updates and automated notifications ensure that all team members, regardless of their location, are informed of the project's status and their individual responsibilities. AI can also facilitate asynchronous collaboration by intelligently routing information and tasks, ensuring that team members can contribute effectively even when they are not online at the same time. This makes it easier to manage global teams and leverage talent from anywhere in the world.

The impact of AI on task management is not limited to generic projects; it can be tailored to specific industry needs. For example, in the construction industry, AI can optimize the scheduling of trades, manage the delivery of materials, and monitor site safety compliance, all of which are critical tasks. In healthcare, AI can assist in patient flow management, appointment scheduling, and resource allocation within hospitals, ensuring that necessary tasks are handled efficiently and that patient care is not compromised. The adaptability of AI allows it to be applied across a broad spectrum of project types and industries, providing tangible benefits wherever complex tasks need to be managed and executed effectively.

However, successfully implementing AI in task management requires careful consideration of several factors. Firstly, data quality is paramount. The AI systems are only as good as the data they receive. Inaccurate, incomplete, or outdated information about team member availability, skill sets, or task statuses will lead to suboptimal recommendations and assignments. Therefore, organizations must invest in robust data management practices and ensure that information is continuously updated and validated. Secondly, the human element remains crucial. AI should be viewed as a tool to augment, not replace, human project managers. Project managers are still needed for strategic decision-making, complex problem-solving, stakeholder management, and fostering team cohesion. The AI provides insights and automates processes, but the ultimate direction and nuanced judgment still reside with human leadership. Finally, ethical considerations must be addressed. Ensuring fairness in task assignment, avoiding bias in AI algorithms, and maintaining transparency in how AI makes decisions are critical for building trust and ensuring that AI is used responsibly. For instance, an AI that disproportionately assigns less desirable tasks to specific team members, even if based on historical data that reflects past biases, would be detrimental to team morale and equitable work distribution. Continuous monitoring and refinement of AI models are necessary to mitigate such risks.

In summary, AI-driven task management and automation represent a significant leap forward in project execution. By enabling intelligent task assignment, real-time progress tracking, automated reporting and reminders, and proactive bottleneck identification, AI empowers project managers to optimize workflows, enhance team productivity, and reduce the administrative burden. This allows teams to focus on higher-value, more creative, and more strategic work, leading to increased job satisfaction, improved project outcomes, and a more agile, efficient approach to project delivery. The intelligent management of tasks, supported by AI, is thus becoming a critical differentiator for organizations seeking to excel in today's competitive landscape.

The landscape of project management is one perpetually fraught with uncertainty. While meticulous planning and skilled execution form the bedrock of successful project delivery, the inherent unpredictability of dynamic environments means that risks are not just possibilities, but inevitabilities. Historically, project managers have relied on experience, intuition, and structured risk assessment frameworks, such as SWOT (**S**trengths, **W**eaknesses, **O**pportunities, and **T**hreats) analyses, FMEA (Failure Mode and Effects Analysis), and risk registers, to identify, evaluate, and plan for potential threats. These methods, while valuable, are often retrospective or based on extrapolating known variables. They excel at cataloging what is known or can be reasonably anticipated. However, they can struggle with the emergent, the subtle, and the complex interdependencies that characterize modern projects, especially those that are heavily reliant on technology or operate in rapidly evolving markets.

This is where the transformative power of AI, specifically through predictive analytics, revolutionizes risk mitigation. Imagine a project manager no longer solely reliant on the compiled wisdom of past projects or the immediate observations of the current team, but armed with an intelligent foresight capable of scanning vast datasets, discerning subtle correlations, and forecasting potential derailments long before they manifest as tangible problems. Predictive analytics, at its core, leverages sophisticated algorithms to examine historical and real-time data, identifying patterns and trends that correlate with specific outcomes, including project risks. This shifts risk management from a purely reactive discipline, where we respond to issues after they arise, to a proactive, strategic function that anticipates and neutralizes threats before they affect project timelines, budgets, or quality.

The process begins with data ingestion. A comprehensive AI-driven risk mitigation system requires access to a diverse array of data points. This includes not only internal project data, such as task durations, resource allocation, budget spend, team performance metrics, and change requests, but also external factors that can profoundly influence project outcomes. These external factors can range from market trends, economic indicators, and competitor activities to supplier performance, regulatory changes, and even meteorological forecasts for construction projects. The AI then employs a variety of analytical techniques, including machine learning models like regression analysis, time-series forecasting, classification algorithms, and anomaly detection, to sift through this data deluge.

Consider, for example, predicting project delays. A traditional approach might involve a project manager noting that similar tasks in past projects often ran over by 15%. An AI system, however, would analyze a far richer tapestry of contributing factors. It could correlate the current task's complexity, the experience level and current workload of the assigned team member, the availability of critical dependencies, such as a third-party software component release, and even external factors, such as known supply chain disruptions for required hardware. If the AI detects a confluence of these indicators, for instance, a highly complex task assigned to a team member with a history of being overloaded, coupled with recent news of a shipping delay for a vital component, it can generate a high-probability alert for a potential delay. This alert is not a vague warning; it can be quantified, providing an estimated delay duration and identifying the specific contributing factors.

Similarly, budget overruns can be anticipated with greater accuracy. Beyond simply tracking current expenditure against budget, AI can analyze patterns in scope creep, the cost impact of change requests, historical pricing trends for procured goods and services, and team productivity fluctuations. If the AI observes a recurring pattern in which a specific type of feature request consistently leads to significant unplanned development costs and extended timelines, it can flag future similar requests as high-risk for budget impact. This allows project managers to engage in more informed discussions with stakeholders, perhaps suggesting alternative, less costly solutions or establishing stricter change-control processes for specific types of requests. The AI acts as an intelligent sentinel, highlighting deviations from expected financial trajectories long before they become critical.

Technical failures, a common pitfall in technology-centric projects, can also be preemptively addressed. By analyzing code repositories, bug-tracking systems, performance-monitoring logs, and even developer activity patterns, AI can identify early warning signs of potential technical issues. For instance, an increase in code complexity in a particular module, a rising trend in unit test failures, or a decrease in code commit frequency from a key developer could all be indicators of underlying technical debt or impending instability. An AI could flag a specific module as having a high probability of failure during integration testing, enabling the development team to allocate additional testing resources or proactively refactor the problematic code segment. This mirrors the methodology employed in cybersecurity, where threat intelligence platforms continuously scan for indicators of compromise to prevent breaches; in project management, this translates to indicators of project compromise.

The actionable insights derived from AI-powered predictive analytics are what truly set this approach apart. The AI does not just present a list of potential problems; it provides context and actionable recommendations. For example, if the AI predicts a delay in a critical path activity due to a specific vendor's historically unreliable delivery schedule, the project manager can explore alternative suppliers, expedite existing orders, or adjust the project schedule to buffer the potential impact. If the AI flags a team member as being at high risk of burnout due to an increasing workload and a lack of recent positive feedback, the project manager can intervene by redistributing tasks, offering additional support, or scheduling a crucial one-on-one conversation. This intelligence allows for targeted interventions, optimizing the allocation of the project manager's own limited time and resources towards the most critical risk mitigation efforts.

The analogy to cybersecurity threat intelligence is particularly apt. Just as security teams use AI to analyze network traffic, identify malware signatures, and predict attack vectors, project managers can use AI to analyze project data, identify risk indicators, and predict potential derailments. The goal in both domains is to move from a reactive stance, patching systems after an attack or fixing a project after it has gone off the rails, to a proactive posture of continuous monitoring, early detection, and preemptive defense. This dramatically reduces the likelihood of catastrophic failures and significantly enhances the overall resilience and success rate of projects.

Implementing AI-driven predictive analytics for risk mitigation is not without its challenges. The accuracy of predictions is heavily dependent on the quality, completeness, and relevance of the data fed into the AI models. "Garbage in, garbage out" is a fundamental truth here. Organizations must therefore establish robust data governance frameworks to ensure that project data is consistently captured, accurately recorded, and regularly updated. Furthermore, the AI models themselves require ongoing training, validation, and refinement to adapt to evolving project dynamics and external environments. A model that was effective for projects a year ago might become less reliable as new technologies emerge, market conditions shift, or team compositions change.

Another critical consideration is the interpretation and integration of AI-generated insights into existing project management workflows. The AI is a tool, albeit a sophisticated one. Project managers must develop the skills to understand the outputs of these systems, critically evaluate the predictions, and judiciously decide on the appropriate course of action. This involves not just understanding the statistical probabilities presented by the AI but also applying human judgment, domain expertise, and an understanding of organizational context. For instance, an AI might flag a task with a high probability of delay. However, a seasoned project manager might know that a particular stakeholder is highly motivated to expedite that specific task, potentially mitigating the predicted delay through their influence. The AI provides valuable foresight, but human oversight and decision-making remain paramount.

Moreover, transparency and explainability are crucial for fostering trust and adoption. Project teams and stakeholders need to understand why the AI is flagging a particular risk. "Black box" AI models, whose decision-making processes are opaque, can lead to skepticism and resistance. Therefore, prioritizing AI solutions that offer explainable AI (XAI) capabilities, which can articulate the reasoning behind their predictions, is highly beneficial. This allows project managers to build confidence in the system and effectively communicate the rationale for their risk mitigation strategies to their teams and stakeholders.

The ethical implications warrant careful consideration as well. Predictive models trained on historical data can inadvertently perpetuate existing biases. For example, if past projects have historically assigned more challenging, high-risk tasks to certain demographic groups, an AI trained on this data might continue to do so, even if the underlying reasons are no longer relevant or justifiable. Project managers and AI developers must actively work to identify and mitigate such biases, ensuring that risk assessments and mitigation strategies are equitable and fair.

Despite these considerations, the strategic advantage of AI-powered predictive analytics for risk mitigation is undeniable. It empowers project managers to shift from a defensive posture of reacting to problems to an offensive strategy of anticipating and neutralizing them. This proactive approach not only increases the likelihood of on-time, on-budget delivery and high-quality outcomes but also conserves valuable organizational resources, reduces stress and rework, and fosters a more stable, predictable project environment. By integrating AI-driven foresight into their risk management frameworks, project managers can navigate the inherent uncertainties of their work with a new level of confidence and control, ultimately safeguarding the project and the organization's strategic objectives. This represents a fundamental evolution in how we approach project risk, moving from simply managing threats to intelligently preventing them, a hallmark of truly advanced project management.

The integration of AI into quality assurance (QA) and testing processes represents a paradigm shift, moving these critical functions from traditionally labor-intensive and often reactive activities to proactive, intelligent, and highly efficient operations. In the complex world of software development and product delivery, ensuring quality is not merely a gatekeeping function; it is a fundamental driver of customer satisfaction, brand reputation, and ultimately, commercial success. AI's capacity to process vast amounts of data, identify subtle patterns, and learn from experience makes it an exceptionally potent tool for augmenting and even revolutionizing how we approach quality assurance and testing.

One of the most impactful applications of AI in this domain is through AI-driven automated testing. Traditional automated testing, while a significant improvement over manual efforts, often requires extensive scripting, maintenance, and expert oversight. AI takes this a step further by introducing intelligence and adaptability into the testing lifecycle. Machine learning algorithms can automatically generate test cases by learning from existing code, user behavior, and defect history. This means that instead of a human tester meticulously designing every permutation of a test scenario, an AI can analyze the application's structure, identify critical functionalities and potential edge cases, and then generate optimized test scripts. For instance, AI can analyze user journey data to prioritize testing of the most frequently used or financially critical workflows. It can also intelligently explore an application, autonomously discovering new paths and functionalities that might have been overlooked by human testers or predefined scripts, thereby expanding test coverage with greater efficiency.

Furthermore, AI enhances the maintenance and resilience of automated test suites. As applications evolve, test scripts need to be updated to reflect changes in the user interface, underlying code, or functionality. This maintenance can be a significant bottleneck. AI-powered tools can automatically detect UI changes and adapt test scripts accordingly, minimizing the manual effort required to keep the automation suite up to date. For example, if a button's label or position changes, an AI-assisted tool can identify this discrepancy and suggest, or even automatically implement, the necessary adjustment in the test script, ensuring that the automation remains effective even in a rapidly changing development environment. This adaptability is crucial for Agile and DevOps methodologies, where frequent releases demand a responsive and robust testing infrastructure.

Beyond script generation and maintenance, AI excels in intelligent test execution and prioritization. Not all tests are created equal, and in large, complex applications, running the entire test suite for every minor code change can be prohibitively time-consuming. AI can analyze code changes and predict which parts of the application are most likely to be affected, or where defects are most likely to arise. By correlating code commits with historical defect data, test execution results, and even developer activity, AI can dynamically prioritize test cases with the highest probability of uncovering critical issues. This "intelligent regression testing" enables development teams to receive faster feedback on the impact of their changes, allowing them to address issues earlier in the development cycle, when they are cheaper and easier to fix. Imagine a scenario where a small change is made to a core authentication module. Instead of running thousands of tests, an AI could identify that this change most critically impacts user login, payment processing, and profile management, and thus prioritize those specific test suites, significantly accelerating the feedback loop.

Another revolutionary aspect of AI in QA is its ability to detect anomalies in code and behavior. While traditional static analysis tools can identify known coding errors or stylistic issues, AI can go much deeper by learning the "normal" patterns of code development and application behavior. By analyzing vast repositories of code, commit histories, and runtime logs, AI models can identify subtle deviations that may indicate bugs, security vulnerabilities, or performance degradation, even if these patterns do not match predefined rules. For instance, an AI might detect an unusual increase in the complexity of a specific code module, a sudden surge in memory usage during a particular transaction, or an unexpected deviation in API response times. These anomalies, which might be too subtle or infrequent for human observation, can be flagged by AI as potential indicators of problems, prompting further investigation before they escalate into critical failures. This capability is particularly valuable for identifying zero-day vulnerabilities and emergent bugs arising from the complex interactions among different software components.

Consider the realm of predictive quality analysis. AI can leverage historical data from past projects, bug reports, customer feedback, and production incidents to build models that predict the likelihood of defects in new code or specific features. By analyzing factors such as code complexity, developers' experience levels, code churn, and results from earlier testing phases, AI can highlight areas of the software that are statistically more prone to defects. This allows project managers to proactively allocate resources, focus more intensive testing efforts on these high-risk areas, or even recommend design or code refactoring to mitigate potential quality issues. This predictive capability transforms quality assurance from a reactive measure of "did we find a bug?" to a proactive strategy of "where are we most likely to encounter a bug, and how can we prevent it?"

For example, a project manager overseeing a large e-commerce platform could use AI to predict which product categories or new features are most likely to experience performance issues or bugs upon launch. The AI might identify that features developed by a specific team, or features involving complex integrations with third-party payment gateways, have historically exhibited higher defect rates. Armed with this insight, the project manager can ensure these areas receive more rigorous performance testing, security audits, and even extended beta testing. This targeted approach not only improves the quality of the final deliverable but also optimizes the use of QA resources, ensuring that testing efforts are concentrated where they will have the most impact.

The impact of AI in QA extends to visual testing and user experience. AI can be trained to recognize expected visual elements and layouts on a user interface. It can then automatically compare screenshots or rendered web pages against a baseline or expected state, identifying discrepancies in design, color, font, or the presence of unexpected elements. This is particularly useful for ensuring UI consistency across different devices, browsers, and screen resolutions, a task that is notoriously tedious and error-prone when done manually. AI can also analyze user interaction patterns to identify usability issues that might not be apparent through functional testing alone, such as confusing navigation flows or drop-off points where users frequently abandon the site.

Furthermore, AI can enhance performance and load testing. By analyzing historical performance data and simulating realistic user loads, AI can predict how an application will behave under stress. It can identify bottlenecks, determine optimal resource allocation, and forecast potential failure points before they occur in a production environment. This predictive capability is invaluable for ensuring applications scale effectively and maintain high performance even during peak demand.

Implementing AI in quality assurance and testing requires a strategic approach. Project managers need to identify the most suitable AI tools and platforms that align with their existing development workflows and organizational goals. This often involves integrating AI capabilities into CI/CD pipelines, enabling continuous feedback and automated quality checks throughout the software development lifecycle. The success of AI integration hinges on the availability of high-quality, relevant data. The AI models need to be trained on comprehensive datasets that include code, test results, bug reports, performance metrics, and user feedback. Therefore, establishing robust data governance and collection processes is a prerequisite for harnessing AI's full potential in QA.

A key benefit for project managers is reduced testing cycles and faster time-to-market. By automating repetitive tasks, intelligently prioritizing test cases, and predicting defects, AI significantly shortens the time required for comprehensive testing. This allows development teams to release new features and products more rapidly, gaining a competitive edge. Moreover, by catching defects earlier in the development process, AI-driven QA reduces the cost of fixing bugs, as issues identified during development are considerably less expensive to resolve than those found in production.

Ultimately, the objective of integrating AI into quality assurance and testing is to elevate product reliability and enhance customer satisfaction. Software that is thoroughly tested, performs well, and is free of critical defects delivers a superior user experience. This, in turn, fosters customer loyalty, improves brand perception, and reduces the number of post-launch issues, support tickets, and costly production fixes. By embracing AI, project managers can ensure that quality is not an afterthought but an intrinsic, intelligently managed aspect of the entire product lifecycle, leading to more successful projects and ultimately, more satisfied customers. This proactive, data-driven approach to quality assurance, powered by AI, is becoming an indispensable component of modern project management, driving efficiency and elevating the standard of digital products and services.

The integration of AI into project management, while offering transformative benefits, also brings forth a complex landscape of ethical considerations that demand careful navigation. As project managers increasingly leverage AI for tasks ranging from resource allocation and risk assessment to stakeholder communication and task automation, understanding and proactively addressing these ethical dimensions is paramount. Failing to do so not only risks undermining the trust and fairness within project teams but can also lead to significant reputational damage and unintended negative consequences for the organization and its stakeholders. This subsection delves into these critical ethical challenges, offering guidance on establishing robust frameworks to ensure AI is deployed responsibly, with a keen awareness of its "Gravity of Command."

One of the most pervasive ethical concerns revolves around data privacy and security. AI systems, by their very nature, often require access to vast amounts of data to learn, analyze, and make predictions. In a project management context, this data can include sensitive information about team members, client contracts, financial projections, intellectual property, and strategic plans. Project managers must ensure that the AI tools they employ adhere to stringent data privacy regulations, such as GDPR, CCPA, or similar local laws, depending on their operational geography. This involves implementing robust security measures to protect this data from unauthorized access, breaches, or misuse. Beyond regulatory compliance, there is an ethical imperative to be transparent with individuals whose data is being processed by AI systems. Team members should understand what data is being collected about their performance, how AI is using it for assessments or allocations, and what measures are in place to protect their privacy. This transparency builds trust and mitigates anxieties associated with surveillance or algorithmic judgment. For instance, an AI tool used for predicting team workload might analyze communication patterns, task completion times, and even calendar entries. Ethically, this data should be anonymized where possible, access should be strictly controlled, and the purpose of its collection and analysis must be clearly communicated to the team.

A related and highly significant ethical challenge is algorithmic bias in decision-making. AI models learn from the data they are trained on. If this historical data reflects existing societal biases, discriminatory practices, or past inequities, the AI will inevitably perpetuate and potentially amplify them in its decision-making. In project management, this can manifest in several detrimental ways. Consider AI-driven resource allocation: an AI might inadvertently favor certain demographic groups for high-profile assignments or promotions if the historical data shows a pattern of such favoritism, even if unintentional. Similarly, AI used for risk assessment could disproportionately flag projects associated with underrepresented teams or communities if past performance data, influenced by systemic disadvantages, suggests a higher risk. This can lead to unfair opportunities, hinder diversity and inclusion efforts, and create an inequitable work environment.

To combat algorithmic bias, project managers must adopt a multi-pronged approach. Firstly, data curation and auditing are crucial. Before deploying any AI system, the training data should be meticulously scrutinized for any inherent biases. This might involve using statistical techniques to identify disparities in representation or outcomes. Secondly, bias detection and mitigation techniques should be integrated into the AI development and deployment process. This can involve using fairness-aware algorithms designed to produce equitable outcomes across groups, or implementing post-processing steps to adjust AI outputs to ensure fairness. Thirdly, human oversight and intervention remain indispensable. AI should be viewed as an augmentation tool, not a replacement for human judgment. Project managers must retain the authority and responsibility to review, challenge, and override AI-generated recommendations, particularly in high-stakes decisions. For example, if an AI suggests reassigning a team member from a critical project based on predicted performance metrics, a human manager should consider the individual's skills, career development goals, and any potential external factors before making a final decision. The "Gravity of Command" here emphasizes that while AI can provide data-driven insights, the ultimate ethical responsibility for fair resource allocation and opportunity rests with the human leader.

Transparency and explainability in AI-assisted project management are also critical ethical imperatives. Many advanced AI models, particularly deep learning systems, operate as "black boxes," making it difficult to understand precisely how they arrive at a particular decision or recommendation. This lack of transparency can erode trust among team members and stakeholders. If an AI flags a project for high risk or recommends a specific course of action, individuals affected by this decision have a right to understand the rationale behind it. Without explainability, it becomes challenging to identify and correct errors, biases, or flawed logic within the AI system. This can lead to decisions made based on inscrutable algorithmic processes, fostering skepticism and resistance.

Project managers should prioritize AI solutions that offer a degree of explainability, often referred to as Explainable AI (XAI). XAI techniques aim to make AI decisions understandable to humans, providing insights into the factors that influenced an outcome. For instance, if an AI recommends a particular project schedule, XAI could highlight the critical path activities, resource constraints, and historical dependencies that led to that recommendation. In the context of performance evaluation, XAI could detail which specific metrics contributed most significantly to an AI-generated assessment. Even when complete explainability is not feasible, project managers must strive for interpretability, ensuring that the general principles and logic underpinning the AI's operation are understood. This might involve maintaining clear documentation about the AI's intended use, its limitations, and the types of data it analyzes. Communicating these aspects openly to project teams and stakeholders is essential for building confidence in the AI-driven processes. The "Gravity of Command" requires leaders not just to accept AI outputs but to understand and be able to articulate their basis, ensuring accountability for decisions, even when supported by AI.

The ethical implications extend to the impact of AI on the project workforce. As AI automates certain tasks, concerns arise about job displacement and the need for upskilling. While AI can free up human project managers from repetitive administrative duties, allowing them to focus on more strategic and creative aspects of their roles, it is essential to manage this transition ethically. Organizations have a responsibility to support their employees through these changes. This includes investing in training and development programs to equip project managers and team members with the new skills required to work alongside AI systems, or to transition into new roles. The goal should be to augment human capabilities, not simply to replace them.

Furthermore, the increasing reliance on AI can alter the dynamics of human interaction and collaboration within project teams. If AI takes over communication scheduling, task delegation, or even performance feedback, there is a risk of reducing opportunities for direct human interaction, empathy, and nuanced understanding, which are vital for team cohesion and morale. Project managers must consciously foster a collaborative environment that balances AI efficiency with meaningful human connection. This might involve designating certain AI-handled tasks as optional, encouraging team members to discuss AI-generated suggestions face-to-face, or ensuring that AI tools are used to facilitate, rather than replace, human dialogue. The "Gravity of Command" here underscores the leader's role in shaping a work culture where technology serves to enhance human relationships and collaboration, not diminish them.

Establishing ethical frameworks and guidelines for AI deployment is no longer optional; it is a necessity for responsible project management. These frameworks should provide a clear set of principles and operational procedures that guide the selection, implementation, and ongoing use of AI technologies. Key elements of such a framework typically include:

- **Purpose and Alignment:** Clearly define the intended purpose of AI use in project management and ensure it aligns with the organization's values, strategic objectives, and ethical commitments. AI should serve human goals and enhance project outcomes, not merely for the sake of technological adoption.

- **Fairness and Equity:** Develop guidelines to ensure that AI systems do not discriminate against any individuals or groups. This includes mechanisms for bias detection and mitigation, as well as regular auditing of AI outputs.
- **Transparency and Accountability:** Establish protocols for transparency into AI use, data processing, and decision-making. Define clear lines of accountability for AI system performance and any adverse outcomes. This means that if an AI makes a flawed recommendation, it must be clear who is responsible for reviewing and acting upon it.
- **Privacy and Security:** Implement robust data governance policies and security measures to protect sensitive project and personal data accessed by AI systems, ensuring compliance with all relevant regulations.
- **Human Oversight:** Mandate appropriate levels of human oversight and intervention for AI-assisted decision-making, particularly in critical areas affecting individuals or significant project outcomes.
- **Continuous Monitoring and Improvement:** Establish processes for ongoing monitoring of AI system performance, ethical implications, and potential unintended consequences. This includes mechanisms for feedback, error reporting, and iterative improvement of AI models and deployment strategies.

Implementing these ethical frameworks requires leadership commitment and active engagement from all levels of the project team. It necessitates ongoing training, open communication, and a culture that prioritizes responsible innovation. The "Gravity of Command" implies that project managers, as leaders, bear the primary responsibility for championing and enforcing these ethical principles. They must act as ethical stewards, ensuring that the power of AI is harnessed for good and fostering trust, fairness, and positive outcomes across all their projects.

Ultimately, the ethical integration of AI in project management is about more than just compliance; it is about building a future where technology augments human potential in a way that is both effective and morally sound. It requires a deliberate and continuous effort to balance the pursuit of efficiency and innovation with the fundamental principles of human dignity, fairness, and accountability. By proactively addressing these ethical considerations, project managers can unlock AI's true potential, ensuring it serves as a powerful ally in delivering successful projects while upholding the highest standards of integrity and responsibility. This mindful approach ensures that the "Gravity of Command" is wielded not just for project success, but for the creation of ethical and sustainable work environments.

4. Earth's Grounding Command

Hiring Paradox

"The algorithm gave him a 98% match score, Bob. He's a safe bet."

Bob looked at the resume on the table. It was perfect—a decade of backend development, PMP certification, and solid references. The AI recruiting tool had highlighted every keyword: Java, SQL, Agile, and Waterfall. It was the kind of resume Bob had hired a hundred times.

"He's safe," Alice countered, sliding a different resume across the table. "But the AI gave this one a 60%. Why? Because she has an undergraduate degree in Philosophy, and her 'years of experience' do not fit the model. But look closer."

Bob squinted. Self-taught Python. Top 5% on Kaggle for a natural language processing challenge. A side project building a chatbot for a local non-profit.

"The AI is hiring for history," Alice said. "It looks at what they have done. I want to hire for trajectory—what they can do. The first candidate is a rock; solid, but immovable. The second candidate is water; fluid, adaptable, and she goes where the cracks are."

*****

The AI Blind Spot: Hiring for Keywords vs. Curiosity

In the age of AI, we face a dangerous temptation: letting the machine filter our talent based on the past. AI recruiting tools are fantastic at matching keywords to job descriptions. They ensure a candidate checks the boxes.

But in a world where the "boxes" change every six months, hiring for compliance is a death sentence.

The Paradox: The candidate who perfectly matches your current job description is often the least equipped to handle your future challenges. Why? Because they have spent ten years perfecting a skill set that AI is about to automate.

*****

The Interview: Testing for the "Un-Googleable"

Bob decided to interview both.

Candidate A (The 98% Match): Bob asked him about a database failure. The candidate gave a textbook answer. He knew the syntax. He knew the protocol.

- **The Red Flag:** When Bob asked, "What would you do if the standard protocol failed and the documentation was offline?" the candidate froze. He was an executor, not a solver.

Candidate B (The 60% Match): Alice asked her about a Python framework she had never used.

- **The Interaction:** She did not fake it. She said, "I do not know that syntax yet. But it looks similar to this library I used last week. Give me five minutes with the documentation, and I will write a prototype." And she did in the interview.

The Leadership Lesson: AI can screen for competence (what they know). Only a human leader can screen for curiosity (how they learn). In "Gravity of Command," we prioritize interaction over algorithm. We look for the "spark"—the ability to navigate ambiguity, to ask the right questions, and to be comfortable not knowing the answer immediately.

*****

Gravity Pulls: The "Turing Test" for Hiring

Do not let the AI make the final call.

- **The "Why" Test:** AI hires people who know what to do. You need to hire people who understand *why* we do what we do. In the interview, ask "Why?" five times. The candidate who gets annoyed is a red flag. The candidate who gets excited is your hire.
- **The "Teach Me" Moment:** Ask the candidate to teach you something new in five minutes. You are not testing their knowledge; you are testing their communication and their passion. AI cannot fake passion.

*****

Summary of the Shift

Bob hired Candidate B. Six months later, when the team had to pivot from a legacy database to a vector database for AI, Candidate A would have been lost. Candidate B led the migration.

The Rule: Use AI to filter the noise, but use your gut and the interview to find the signal. Hire for the learning curve, not the experience curve.

This is the essence of Earth's Grounding Command. Just as Earth provides the solid ground for life to flourish, a leader must provide the fertile soil for a new kind of team. In the age of AI, you are not just a manager of tasks; you are a curator of talent.

The New Talent Stack: Beyond the Resume

The first challenge for Gravity Leaders is realizing that the "perfect candidate" has changed. As AI automates routine coding and testing, the value shifts from execution (doing what you are told) to orchestration (guiding the machine).

1. The Technical Shift: From Generalist to Specialist. We still need coders, but the roles are evolving. We need:

- Data Engineers: To build the pipelines that feed the AI.
- MLOps Specialists: To keep the models running in production.

- Prompt Engineers: Who understand how to talk to LLMs to get the best output.

2. The Soft Skill Revolution. More importantly, the “soft” skills are becoming the “hard” requirements.

- Analytical Thinking: Can they look at an AI prediction and ask, “Does this make sense?” or do they unquestioningly trust the black box?
- **Adaptability:** The tools we use today will be obsolete in six months. Can they unlearn and relearn?
- Ethical Awareness: Can they spot bias in a dataset before it becomes a PR disaster?

*****

Onboarding: Planting the Seeds

Once you hire the resources, how do you ground them? In traditional projects, onboarding meant “here is your laptop and the documentation.” In AI projects, that is a recipe for failure.

The “Gravity” Onboarding Model:

- **Phase 1 (Week 1): The Human Connection.** Before your new team member touches a line of code, they meet the stakeholders. They need to understand the business context (The North Star) so they do not build technically perfect but useless models.
- **Phase 2 (Weeks 2-4): The Sandbox.** Give those new hires a “safe” project. Let them break things. Assign a mentor (a “Buddy”) who can guide them through the unwritten rules of the team’s data culture.
- **Phase 3 (Month 2+): Ethical Stewardship.** Explicitly train the new hires on your AI ethics guidelines. They need to know that raising a red flag about bias is not “slowing down”—it is part of their job.

*****

Gravity Pulls: The Learning Culture

"The team that learns the fastest wins." You cannot just tell people to learn. You must create the space for it.

- The 80/20 Rule: Encourage your team to spend 80% of their time on delivery and 20% on experimentation (e.g., "Innovation Sprints" or Hackathons).
- **Fail Forward:** When an experiment fails, celebrate the insight. "We learned that this algorithm does not work for this data type. That is a win."

*****

The AI Motivation Engine: From Monitoring to Empowering

"You cannot algorithm your way to morale," Bob once grumbled. But Alice showed him that AI, when used correctly, is not Big Brother—it is a force multiplier for motivation.

1. The Drudgery Killer (Restoring Autonomy) Nothing kills motivation faster than "gravity drag"—the heavy, repetitive tasks.

The Strategy: Do not implement AI to "cut headcount." Implement it to "upgrade roles." Use AI to handle logs, reports, and routine tickets so your humans can get back to creative engineering.

2. The Invisible Spotlight (Data-Driven Recognition) We often celebrate the "firefighters" who fix crises but ignore the "fire marshals" who prevent them.

The Strategy: Use project analytics to see who is doing the invisible work—updating documentation, fixing minor bugs, helping peers. When you publicly praise someone for "cleaning the data pipeline," you signal that stability matters as much as speed.

3. The Personal Trainer (Accelerating Mastery) High performers leave when they stop learning.

The Strategy: Use AI-driven learning platforms to create hyper-personalized growth paths. Encourage your team to use Generative AI as a "sparring partner" to learn new languages or frameworks instantly, accelerating their mastery.

*****

The Mirror and the Megaphone: Feedback in the AI Era

Feedback is the breakfast of champions, but it is often served cold (months late) and soggy (vague). AI gives leaders two new tools:

1. The Mirror (Objective Reflection) Traditional feedback feels personal. "You write bad code." AI feedback is objective. "The static analysis shows your code complexity increased by 15% this sprint." By using tools that analyze code quality or communication patterns, you give employees a "private mirror" to self-correct before you even have to intervene.

2. The Megaphone (Amplifying Wins) Waiting for the annual review to say "good job" is a morale killer.

The Strategy: Configure your tools to generate a "Friday Wins" summary. "This week, the team closed 42 tickets. Special shout-out to Maria for reducing database latency by 5%." Data detects the win; the leader delivers the feeling.

The success of any AI-integrated project hinges fundamentally on the caliber of the team assembled to execute it. Unlike traditional projects where skill sets might be more standardized, AI initiatives demand a unique blend of technical prowess, analytical acumen, and a high degree of adaptability. Building such a team requires a strategic, nuanced approach to both recruitment and onboarding, ensuring that the right talent is not only identified but also seamlessly integrated to foster immediate and sustained contributions. This subsection delves into the critical aspects of cultivating high-performing AI-centric teams, focusing on identifying essential skills, effective recruitment strategies, and robust onboarding processes that lay the foundation for success.

Identifying the core competencies required for AI projects is the crucial first step. These can be broadly categorized into technical skills and essential soft skills. On the technical front, the demand for roles such as Data Scientists and Machine Learning Engineers is self-evident. Data Scientists are adept at exploring, cleaning, and transforming data, uncovering patterns, and formulating hypotheses. They possess strong statistical knowledge, proficiency in programming languages such as Python and R, and experience with various data analysis and visualization tools. Machine Learning Engineers, on the other hand, focus on building, deploying, and scaling machine learning models. They require a deep understanding of algorithms and software engineering principles, as well as experience with ML frameworks such as TensorFlow, PyTorch, or scikit-learn.

Beyond these specialized roles, there is a growing need for AI/ML Operations (MLOps) specialists, who bridge the gap between development and operations, ensuring the smooth deployment, monitoring, and maintenance of ML models in production environments. This often involves expertise in cloud platforms, containerization technologies like Docker and Kubernetes, and CI/CD pipelines. Furthermore, individuals with strong data engineering backgrounds are indispensable for building and maintaining the robust data pipelines and infrastructure that AI models rely upon. Their skills in database management, ETL processes, and distributed data systems ensure that high-quality, accessible data is available for AI development and deployment.

However, technical skills alone are insufficient. The dynamic nature of AI projects and the inherent uncertainties they entail require a strong suite of soft skills. Analytical thinking is paramount; team members must be able to break down complex problems, critically evaluate information, and draw logical conclusions, even when dealing with incomplete or ambiguous data. This is coupled with problem-solving aptitude—the ability to devise creative and practical solutions to novel challenges that AI development inevitably presents.

Adaptability and a growth mindset are equally critical. The field of AI is evolving at an unprecedented pace, with new research, tools, and techniques constantly emerging. Team members must be eager to learn, embrace new methodologies, and pivot their strategies when necessary. This also includes a willingness to experiment and iterate, understanding that failures are often learning opportunities on the path to success. Effective communication is vital, particularly for explaining complex AI concepts to non-technical stakeholders, collaborating with diverse team members, and documenting processes and findings clearly. Collaboration itself—the ability to work effectively in cross-functional teams, share knowledge, and provide constructive feedback—is the bedrock of any high-performing team. Finally, ethical awareness and a strong sense of responsibility are increasingly important as AI's impact becomes more profound, ensuring that projects are developed and deployed in a fair, transparent, and beneficial manner.

Recruiting for these specialized roles presents unique challenges. The demand for AI talent often outstrips supply, leading to intense competition. To attract the best candidates, organizations must move beyond traditional recruitment channels and adopt more targeted and appealing strategies. Firstly, developing a compelling employer brand that highlights the organization's commitment to innovation, its cutting-edge AI projects, and its supportive work environment is crucial. This can be achieved through thought leadership content, participation in industry conferences, and showcasing successful AI initiatives. Secondly, leveraging niche professional networks and communities, such as GitHub, Kaggle, and specialized AI forums, can provide access to a pool of highly skilled individuals actively engaged in the AI ecosystem. Participating in or sponsoring hackathons and AI competitions can also be an effective way to identify and engage with emerging talent.

When crafting job descriptions, it is essential to be precise about the required technical skills, project context, and the expected impact of the role. While listing every possible tool and technology can be overwhelming, focusing on core competencies and frameworks relevant to the specific project is more effective. Equally important is clearly articulating the desired soft skills and cultural fit. For senior roles, emphasize experience in leading AI initiatives, managing complex projects, and mentoring junior team members.

The recruitment process itself needs to be streamlined and efficient. AI candidates are often in high demand and may be evaluating multiple opportunities. Long, convoluted hiring processes can lead valuable candidates to accept offers elsewhere. This means ensuring that application reviews are prompt, interview stages are well-defined and relevant, and feedback is provided on time. Interviews should assess not only technical proficiency through coding challenges or case studies but also problem-solving skills, communication abilities, and cultural alignment. For technical roles, involving existing AI team members in the interview process can provide valuable insights into a candidate's technical depth and their potential fit within the team dynamic. A practical approach, such as a take-home assignment or a pair-programming exercise, can offer a more realistic assessment of a candidate's skills than purely theoretical questions. It is also beneficial to assess a candidate's motivation and their understanding of the ethical considerations pertinent to AI development, as this reflects their maturity and foresight.

Onboarding new talent into AI projects requires a structured yet flexible approach to ensure rapid integration and productivity. The primary goal is to help new team members understand the project's objectives, their specific role and responsibilities, the existing technical infrastructure, and the team's working norms and culture. A comprehensive onboarding plan typically spans several weeks and includes distinct phases.

The initial phase, often the first week, focuses on essential administrative tasks, introductions, and foundational knowledge. This includes setting up necessary hardware and software access, completing HR paperwork, and providing an overview of the company's mission, vision, and values. Crucially, this period should involve formal introductions to the core project team, key stakeholders, and relevant departments. New hires should receive an in-depth briefing on the project's goals, current status, challenges, and the overall AI strategy. Providing access to relevant documentation—project charters, technical specifications, past reports, and code repositories—is essential for self-study. A designated "buddy" or mentor from the existing team can be incredibly valuable during this time, acting as a go-to person for informal questions, navigating company culture, and facilitating early connections.

The second phase, typically weeks two through four, delves deeper into the technical and operational aspects of the project. This involves understanding the specific AI models, algorithms, and tools being used, the data pipelines and architecture, and the deployment and monitoring processes. Hands-on tasks, starting with smaller, well-defined assignments, are key to building confidence and practical experience. For instance, a new data scientist might be tasked with data exploration and initial feature engineering on a specific dataset, or a machine learning engineer might be assigned to adapt an existing script or set up a new experiment environment. Regular check-ins with their manager and mentor are vital for providing feedback, addressing roadblocks, and adjusting the onboarding pace as needed. This is also the time to introduce the team's collaborative tools and methodologies, such as version control workflows, agile sprint cadences, and communication protocols.

As onboarding progresses over the subsequent months, the team shifts its focus toward increasing autonomy and contribution. The team gradually assigns the new member more complex tasks and integrates them into the core project workflows. The team actively creates opportunities for the new member to present their work, participate in design discussions, and contribute to problem-solving sessions.

The organization encourages continued learning and development by providing access to training resources, workshops, and opportunities to explore new AI technologies relevant to the project. Leaders clearly communicate performance expectations and establish regular feedback loops to monitor progress and identify areas for growth.

A critical aspect of onboarding for AI projects is fostering an understanding of the ethical implications and responsible AI practices. New hires should be educated on the company's AI ethics guidelines, data privacy policies, and any specific protocols related to bias detection and mitigation, transparency, and accountability. Discussions around potential societal impacts of the AI systems being developed should be encouraged, embedding a sense of responsibility from the outset. This proactive approach ensures that ethical considerations are not an afterthought but an integral part of the development lifecycle.

Furthermore, the onboarding process should be designed to foster a sense of belonging and team cohesion. AI projects can sometimes be highly specialized, leading to siloed work. Encouraging cross-functional learning, team-building activities, and knowledge-sharing sessions can help bridge these divides. New hires should feel empowered to ask questions, challenge assumptions, and contribute their unique perspectives. The goal is not just to train individuals on specific tasks but to integrate them into a collaborative ecosystem where diverse skills and viewpoints are valued and contribute to the collective success of the AI initiative. Ultimately, a well-executed recruitment and onboarding strategy for AI projects is an investment that pays dividends in terms of team performance, innovation, and the successful realization of AI-driven objectives. It transforms individual talent into a cohesive, skilled, and motivated unit capable of navigating the complexities of AI and delivering impactful results.

The relentless pace of advances in artificial intelligence demands a deep commitment to continuous learning and adaptation within teams. What is state-of-the-art today can become commonplace tomorrow, and therefore, a team's ability to not just keep up but to proactively anticipate and integrate new developments is a critical determinant of its long-term success. Cultivating a culture where learning is not an occasional event but an embedded, ongoing practice requires deliberate leadership and the creation of an environment that actively encourages curiosity, experimentation, and resilience. This is not merely about acquiring new technical skills, although that is a significant component; it is about fostering a collective mindset that embraces change as an opportunity rather than a threat and views challenges as catalysts for innovation.

To foster this culture, leaders must first champion the value of learning. This begins with setting clear expectations that professional development and the exploration of new AI frontiers are not optional extras but integral to the team's performance and individual growth. This can be communicated through regular team meetings, one-on-one discussions, and by visibly allocating time and budget for learning initiatives. When team members see that their organization and leadership genuinely invest in their continuous upskilling, they are more likely to reciprocate with their own proactive engagement. Leaders can also model this behavior themselves, openly sharing their own learning journeys, the resources they find valuable, and the insights gained from exploring new AI concepts. This vulnerability and transparency can demystify the learning process and encourage others to step outside their comfort zones.

Creating psychological safety is another cornerstone of a learning-oriented culture. Team members need to feel secure in the knowledge that they can explore new ideas, ask "naive" questions, and even experiment with approaches that might not yield immediate success without fear of reprisal or judgment. This is particularly crucial in AI, where experimentation is often necessary to discover the most effective algorithms, data preprocessing techniques, and model architectures. A culture that penalizes failure stifles innovation. Instead, teams should be encouraged to view experiments as data-gathering exercises, where even unsuccessful attempts provide valuable lessons that inform future endeavors. This requires leaders to reframe the narrative around failure, celebrating the learning derived from setbacks rather than focusing solely on the outcome. Post-mortems, for instance, should be framed as "learning reviews" that focus on understanding what happened, why it happened, and what can be done differently next time, rather than assigning blame.

Experimentation needs to be a tangible part of the team's workflow. This can manifest in various ways. Allocating "innovation time" or "learning sprints" where team members are encouraged to work on side projects, explore new tools, or delve into research papers can be highly effective. This does not have to be a completely unstructured free-for-all; it can be guided by broader strategic objectives or emerging trends identified by leadership. For example, a team might collectively decide to spend a week exploring the capabilities of a new generative AI model relevant to their domain, culminating in a presentation of potential applications. Another approach is to integrate small-scale experiments directly into ongoing project sprints. This could involve dedicating a sprint to testing a novel feature, evaluating an alternative algorithm, or optimizing a specific part of the data pipeline. The key is to make experimentation a regular, accepted practice, rather than a rare, memorable event.

Knowledge sharing is the vital connective tissue that binds individual learning into collective growth. A culture of continuous learning thrives when knowledge is not hoarded but actively disseminated. Leaders can facilitate this by establishing diverse forums for knowledge exchange. Regular "lunch and learn" sessions, where team members can share insights from recent articles, conference talks, or personal projects, are excellent for informal learning. "Tech talks" or "deep dives" can be organized for more in-depth explorations of specific topics, perhaps led by a team member who has recently acquired expertise in a particular area. Establishing internal wikis, shared document repositories, or dedicated Slack channels for discussing new research papers or tools can also create persistent knowledge bases. Encouraging team members to present their findings and learnings to the broader group, even if the insights are preliminary, reinforces the value of sharing and can spark collaborative problem-solving.

Beyond formal and informal sharing sessions, mentorship and peer learning are invaluable. Pairing experienced team members with those who are newer to AI or to specific technologies can accelerate individual development and foster cross-pollination of ideas. This does not always need to be a formal, top-down assignment; encouraging organic mentorship relationships can be equally powerful. Leaders can facilitate this by creating opportunities for cross-functional interaction and by recognizing and rewarding individuals who actively mentor their colleagues. Team-based problem-solving, where complex challenges are tackled collaboratively rather than by individuals in isolation, also inherently promotes learning as team members draw upon each other's diverse skill sets and perspectives.

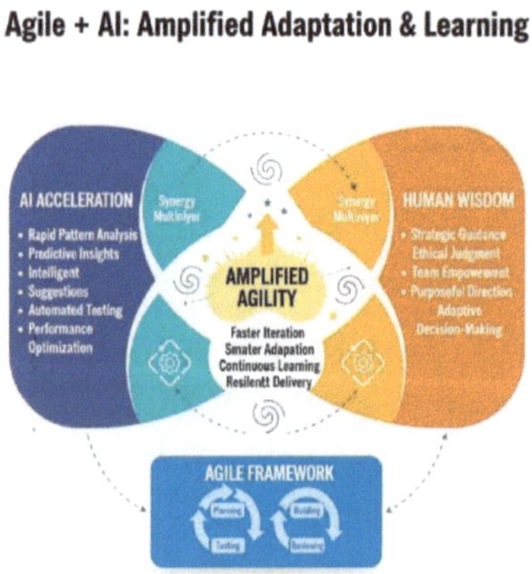

The iterative nature of AI development itself provides a natural framework for continuous learning and adaptation. Agile methodologies, which are widely adopted in software development, are particularly well-suited for AI projects. The cyclical process of planning, building, testing, and reviewing allows teams to constantly reassess their approach, incorporate feedback, and adapt to new information. Leaders should ensure these agile principles are not merely followed mechanically but embraced with a spirit of continuous improvement. Regular retrospectives, conducted with honesty and a focus on actionable insights, are critical for identifying what is working well, what could be improved, and how the team can better adapt to evolving project requirements or unexpected technical hurdles.

Adaptation to evolving project requirements or unexpected challenges is a hallmark of high-performing AI teams. This requires not only the willingness to change course but also the agility to do so effectively. Leaders play a crucial role in managing this adaptation. When unforeseen issues arise—a model's performance degrades unexpectedly, a new regulatory requirement emerges, or a promising new AI technique becomes available—leaders must guide the team through the necessary adjustments. This might involve re-prioritizing tasks, reallocating resources, or even fundamentally rethinking the project's architecture. The key is to maintain a clear vision while remaining flexible in the execution. This also involves fostering a degree of resilience within the team. AI projects can be characterized by periods of intense work followed by moments of recalibration. A resilient team can weather these fluctuations, maintain motivation, and quickly pivot back to productive work when necessary.

To stay at the cutting edge, teams need access to relevant information and training. This goes beyond internal knowledge sharing. Leaders should actively encourage and support team members in attending industry conferences, participating in online courses and certifications, subscribing to relevant academic journals and technical blogs, and engaging with professional communities. For instance, a team working on natural language processing might encourage its members to follow leading AI researchers on social media, subscribe to NLP newsletters, and attend virtual workshops on the latest transformer architectures. The investment in such external learning opportunities not only equips individuals with new skills but also exposes them to broader trends and potential future directions, allowing the team to remain agile and responsive to emerging opportunities.

Moreover, fostering a culture of continuous learning involves proactively evolving the team's skill set, rather than reacting to it. This means anticipating future needs based on industry trends, emerging technologies, and the organization's strategic goals. Leaders should work with their teams to identify skill gaps that may exist today or emerge in the near future. This could involve identifying a need for expertise in a new area, such as explainable AI (XAI) or ethical AI governance, and then developing a plan to either train existing team members or recruit individuals with those skills. This forward-looking approach ensures that the team is not constantly playing catch-up but is positioned to leverage new advancements effectively.

The integration of AI tools and platforms into everyday workflows also requires a learning mindset. As new AI-powered assistants, code generators, or data analysis tools become available, teams should be encouraged to explore their utility and incorporate them where appropriate. This requires an open mind and a willingness to experiment with new ways of working. For example, integrating AI-powered code completion tools can significantly boost developer productivity, but it requires developers to learn how to use these tools effectively and understand their limitations. Similarly, using AI for automated data labeling or feature engineering requires a learning curve. Leaders should facilitate this by providing access to these tools, offering basic training, and encouraging team members to share best practices and tips.

Finally, the development of AI systems is inherently a journey of continuous refinement and improvement. This principle must extend to the team's processes and methodologies. Regular reviews of team dynamics, communication channels, and workflow efficiency are essential. Are the current project management tools serving the team well? Could the code review process be more efficient? Are there bottlenecks in the deployment pipeline that can be addressed? By fostering a culture where team members feel empowered to identify and propose improvements to internal processes, leaders can ensure the team is continuously learning and adapting, becoming more effective and efficient over time. This commitment to internal optimization, alongside the focus on technical skills and new AI advancements, creates a truly high-performing, adaptive, and future-ready AI-integrated team.

The cultivation of truly high-performing AI-integrated teams hinges on a leadership philosophy that champions both autonomy and innovation. It is about shifting from a directive, command-and-control model to one that empowers individuals, trusts their capabilities, and provides the fertile ground for creative breakthroughs, especially when leveraging the transformative power of artificial intelligence. This means moving beyond simply assigning tasks and towards enabling team members to become proactive owners of their work, encouraging them to explore novel solutions and drive progress independently.

A cornerstone of empowering autonomy is effective delegation. This isn't merely about offloading tasks; it is a strategic act of entrusting responsibility, providing growth opportunities, and demonstrating confidence in an individual's or a subgroup's ability to deliver. For AI-integrated teams, delegation takes on a nuanced dimension. When assigning a project involving, for instance, the development of a predictive maintenance model, a leader might not dictate the specific algorithms to be used or the exact data preprocessing steps. Instead, they would clearly define the desired outcome—a model that accurately predicts equipment failure with a specified confidence interval—and the constraints, such as acceptable latency or computational resources. This leaves ample room for team members to exercise their judgment, explore different AI architectures, such as time-series models like LSTMs, and even ensemble methods, experiment with various feature engineering techniques, and select the most appropriate AI tools and libraries. The act of delegation, when done thoughtfully, becomes an explicit invitation for creative problem-solving and encourages individuals to take full ownership of the solution's design and implementation. It signals that their expertise is valued and that their approach to leveraging AI technologies is respected.

Crucially, this delegation must be accompanied by a robust framework of constructive feedback. Feedback, in this context, is not about micromanagement or simply pointing out errors. It is a continuous dialogue aimed at fostering growth and refining approaches. For AI projects, feedback loops are particularly vital. A team member might have developed an initial version of a natural language processing model for sentiment analysis. Constructive feedback would go beyond stating that the accuracy is below par. It would involve a collaborative discussion about *why* it might be underperforming—perhaps the training data lacked sufficient diversity, the chosen embedding techniques were suboptimal, or the model architecture was not suited for capturing nuanced linguistic expressions. This feedback should be specific, actionable, and delivered in a supportive manner, often during regular check-ins or sprint retrospectives. The goal is to guide the individual or team towards self-correction and deeper understanding, enabling them to iterate on their AI solutions more effectively and learn from each adjustment.

The bedrock upon which both autonomy and innovation are built is psychological safety. In an environment where team members feel psychologically safe, they are more likely to take calculated risks, propose unconventional ideas, and voice concerns without fear of negative repercussions. This is paramount in AI development, a field rife with experimentation and inherent uncertainty. Imagine a scenario where a team is tasked with creating an AI system to automate complex customer service queries. A junior member might have an idea to explore a novel reinforcement learning approach, which, while potentially groundbreaking, carries a higher risk of failure or a steeper learning curve. In a psychologically unsafe environment, this individual might suppress their idea, fearing ridicule or being seen as incompetent if the approach does not pan out. Conversely, in a psychologically safe team, this idea would be welcomed, discussed, and potentially explored, perhaps through a small-scale proof-of-concept. Leaders play a pivotal role in fostering this safety by actively encouraging diverse perspectives, refraining from blame when experiments do not yield desired results, and visibly supporting learning from failures. They create an atmosphere where "it didn't work" is a valuable data point, not a personal indictment.

Balancing guidance with autonomy is a delicate but essential leadership act. Leaders of high-performing AI teams are not absent; they are present as facilitators, mentors, and strategic anchors. They provide clear direction on the overarching goals and the business context of AI initiatives, ensuring the team's efforts align with organizational objectives. However, they consciously resist the urge to dictate every step of the implementation process. For instance, when a team is building an AI-powered recommendation engine, the leader would define the desired user experience and key performance indicators, such as click-through and conversion rates, while allowing the team to determine the best collaborative filtering algorithms, appropriate feature selection methods, and deployment strategy. This balance enables individuals to exercise their ingenuity and find the most effective ways to harness AI tools, fostering a sense of ownership and pride in their work. It encourages them to think critically about how best to apply AI to solve the problem at hand, rather than simply executing a prescribed plan.

This empowerment directly translates into enhanced individual motivation and job satisfaction. When team members are given the freedom to explore, experiment, and take ownership of their AI projects, their engagement levels naturally increase. They are more likely to invest discretionary effort, driven by intrinsic motivation rather than external pressure. This can be observed in the enthusiasm with which they tackle complex coding challenges, delve into cutting-edge research papers to find novel AI solutions, or propose innovative ways to optimize model performance. The feeling of having contributed meaningfully to a project, of having applied their skills to solve a challenging problem using advanced AI techniques, is profoundly rewarding. This leads to a more positive work environment, reduced burnout, and a greater likelihood of retaining talented AI professionals who feel valued and are given opportunities for continuous intellectual stimulation.

Furthermore, empowering autonomy unleashes the team's collective intelligence, leading to breakthrough ideas and solutions that significantly advance project goals. AI projects are often multifaceted and require diverse skill sets and perspectives. When individuals are empowered to bring their unique insights and approaches to the table, the team becomes a dynamic engine of innovation. Consider a team developing an AI system for medical image analysis. One member, with expertise in computer vision, might propose a convolutional neural network architecture. Another, with a background in bioinformatics, might suggest incorporating genetic data as an additional input feature, leading to a more robust and personalized diagnostic tool. A third, with an understanding of ethical AI, might highlight the need for bias detection and mitigation strategies early in the development cycle. When autonomy is granted and psychological safety is present, these diverse contributions can converge, leading to solutions that are not only technically sound but also innovative, ethical, and impactful. This collaborative synergy, fueled by empowered individuals, is what distinguishes a good AI team from a truly exceptional one.

To practically implement this empowerment, leaders can introduce specific practices. "Innovation Sprints" or "Hackathons" dedicated to exploring novel AI applications, even if they are tangential to current projects, can provide a structured yet autonomous environment for experimentation. These events allow team members to step away from their day-to-day tasks, collaborate with colleagues they might not typically work with, and apply AI tools in creative ways. The outcomes, whether a functional prototype or a well-researched proposal, can then be evaluated for their potential to be integrated into the broader project roadmap or future initiatives. Similarly, fostering a "maker culture" where team members are encouraged to build and share personal AI projects can be highly beneficial. This provides a low-stakes avenue for learning new AI techniques and tools, and often leads to unexpected discoveries that can be brought back to the team's core work. The leader's role here is to provide resources, remove roadblocks, and champion the spirit of exploration.

Providing constructive feedback also needs to be framed in a way that supports autonomy. Instead of saying, "This model's accuracy is too low," a leader might ask, "What approaches have you considered to improve the model's accuracy, and what are the trade-offs you foresee with each?" This Socratic approach encourages the individual to think critically and arrive at their own solutions, thereby reinforcing their sense of ownership and capability. When feedback is delivered, it should ideally be a two-way conversation, allowing the team member to explain their rationale and offer their own perspectives on potential improvements. This collaborative refinement process is far more effective than a top-down critique in fostering both skill development and sustained motivation.

Furthermore, leaders must be mindful of the tools and resources available to their teams. Empowering autonomy in the context of AI means ensuring team members have access to the computational power, datasets, and advanced AI libraries and frameworks they need. It also involves providing training and guidance on how to utilize these resources effectively. For example, if a team is exploring generative AI for content creation, the leader should ensure they have access to powerful GPUs and are trained on prompt engineering techniques. The leader's role is to facilitate the environment and the resources that enable autonomous exploration and innovation.

Ultimately, empowering autonomy and fostering innovation within AI-integrated teams is not just about achieving project goals more effectively; it is about cultivating a team culture that is resilient, adaptable, and continuously evolving. When individuals feel trusted and empowered to take ownership, they become more invested in the team's and the organization's success. They are more likely to proactively identify opportunities for improvement, propose novel solutions to complex problems, and embrace the inherent challenges of AI development with a spirit of inquiry and determination. This creates a virtuous cycle where empowered individuals drive innovation, leading to breakthroughs that not only meet current objectives but also position the team and the organization for future success in the rapidly advancing landscape of artificial intelligence. The leader's success in this paradigm is measured not just by the team's output but by the growth, engagement, and innovative capacity of each individual within it.

Managing Performance and Providing Constructive Feedback

In the dynamic realm of AI-integrated teams, effective performance management is not merely an administrative task; it is a strategic imperative that underpins sustained success and fosters a culture of continuous improvement. This involves a multifaceted approach, beginning with the foundational establishment of clear expectations. For teams working with artificial intelligence, these expectations must be meticulously defined, encompassing not only the desired business outcomes but also the ethical considerations and the standards of rigor expected in their AI development processes.

When setting objectives for an AI project, such as developing a customer churn prediction model, it's crucial to articulate what "success" looks like. This might include a specific accuracy threshold, a defined reduction in churn rate, the interpretability of the model's predictions, and adherence to data privacy regulations. Leaders must ensure that these objectives are not only communicated but are also understood and agreed upon by all team members, creating a shared vision that guides their efforts. The **SMART framework**—Specific, Measurable, Achievable, Relevant, and Time-bound—remains an invaluable tool, but it must be applied with an AI-specific lens. For instance, "achievable" might involve considering the limitations of available data or the computational resources for training complex models, while "relevant" ensures the AI initiative directly contributes to broader organizational goals.

Defining measurable outcomes is intrinsically linked to setting expectations.

Here is what the data reveals across industries (McKinsey 2025). Understand where your team's function sits in the productivity potential curve:

Function	Potential Lift	Real Results	Why It Works	Your Actionable KPI
Customer Service Operations	30-45% of function costs [McKinsey]	14% ↑ issue resolution/hour 9% ↓ handling time	GenAI helps less-experienced agents mimic expert communication	Reduce cost per contact by 12% in 6 months
Software Engineering	56% coding speed ↑ [GitHub Copilot] 90% Fortune 500 adoption	-	Speed ≠ Quality (maintain code complexity metrics)	Reduce time-to-feature by 40% while maintaining code complexity <15% ↑
Marketing & Sales	Marketing: 5-15% Sales: 3-5% of spend	Michaels Stores: Email personalization 20%→95% 41% ↑ SMS click-through	Personalization at scale	Increase campaign personalization 30%→75% in 90 days

R&D / Research	Months→weeks (drug discovery)	Companies using AI: Phase 3 progression without typical setbacks	Accelerates lead identification	Accelerate lead identification 40% while maintaining clinical success rates.

What this means for your team:

The key is matching potential to your domain AND tying it to a measurable delta within the renewal cycle. Generic 'improve productivity' targets kill adoption. Specific, enforceable targets (with owners and timelines) enable Gravity Leadership."

For AI-integrated teams, this often means identifying Key Performance Indicators (KPIs) that accurately reflect the impact and effectiveness of the AI solutions being developed or deployed. These KPIs can range from technical metrics, such as model accuracy, precision, recall, and $F1$-score for machine learning models, to business-oriented metrics, such as conversion rates, customer satisfaction scores, operational efficiency gains, or cost reductions attributed to the AI system.

Example: A team building an AI-powered chatbot for customer support might track:

- **First-contact resolution rate**
- **Average handling time**
- **Customer satisfaction (CSAT)**
- **Query deflection percentage**

The crucial aspect is that these measures should be communicated transparently and regularly reviewed. Leaders must resist the temptation to focus solely on raw performance numbers without considering the context or the process that generated them. The journey of developing an AI solution is often iterative and experimental, and understanding the performance trajectory, not just the final outcome, provides valuable insights into the team's progress and potential areas for improvement.

The advent of AI itself offers novel avenues for performance tracking, but it is paramount that managers maintain a human-centric approach to interpreting this data and delivering feedback. AI-driven tools can automate the monitoring of code commits, model performance drift, and resource utilization. For instance, sophisticated **MLOps** (Machine Learning Operations) **platforms** can flag when a deployed model's predictions degrade due to changes in input data, signaling a need for retraining or recalibration.

However, these tools are supplements, not replacements, for human judgment. A dip in a model's accuracy, as flagged by an AI system, requires a manager to engage with the team to understand the underlying causes. Is it a data quality issue? Was there a flaw in the feature engineering process? The AI can highlight the problem, but it cannot diagnose the nuanced reasons or deliver the empathetic feedback needed for growth. The manager's role is to synthesize the data from these AI tools with qualitative observations of team dynamics, individual contributions, and the overall project context to provide a holistic performance assessment.

Providing regular, actionable feedback is the linchpin of performance management, especially within AI teams where the technological landscape is constantly evolving. Feedback should be a continuous dialogue, not an annual event. For AI projects, feedback often needs to be technically specific yet accessible. Instead of a vague comment like "the algorithm isn't performing well," a more constructive approach would be:

"*I've noticed the precision in identifying fraudulent transactions has dropped by 10%. Have we explored the impact of recent shifts in transaction patterns, or could there be an opportunity to refine the feature set related to user behavior?*"

This type of feedback is actionable because it points to a specific area of concern and suggests potential avenues for investigation. It also encourages the team member to take ownership of the problem-solving process.

Addressing underperformance requires a delicate balance of support and accountability. When an individual's performance consistently falls short, it's essential to approach the situation with a problem-solving mindset. The first step is a direct and private conversation to understand the root cause. Is it a lack of skills or knowledge related to specific AI tools? Is it a misunderstanding of project requirements?

If a skills gap is identified, the manager should collaborate with the individual to create a **development plan**. This might include targeted training, mentorship, or access to courses on relevant AI technologies, such as deep learning frameworks or natural language processing libraries. For example, if a team member struggles to optimize a neural network's inference speed, the plan might include focused sessions on techniques such as model quantization or pruning. The manager's role is to provide the resources and support necessary for the individual to improve, while also setting clear timelines and checkpoints to monitor progress.

Conversely, recognizing and rewarding high achievers is equally crucial for maintaining team morale and fostering a culture of excellence. This goes beyond mere monetary compensation. Public acknowledgment of exceptional contributions, whether in team meetings or company-wide communications, can be highly motivating. For an AI engineer who develops a novel feature engineering technique that significantly boosts model accuracy, specific praise detailing the impact of their innovation is far more impactful than a generic "good job."

This recognition can also take the form of offering opportunities for professional development, such as attending prestigious AI conferences, leading new research initiatives, or mentoring junior team members. Providing autonomy on challenging, high-visibility projects can also serve as a powerful form of recognition, signaling trust and confidence in their capabilities. For instance, assigning a lead role in developing a cutting-edge generative AI application to a consistently high-performing individual demonstrates that their expertise is valued and that they are being invested in for future leadership roles.

Furthermore, it is essential to differentiate between occasional dips in performance and a persistent pattern of underachievement. An individual might struggle with a particular AI algorithm or a specific phase of the ML lifecycle, but excel in others. Performance management should acknowledge these variations. Acknowledging strengths while addressing weaknesses creates a more balanced and encouraging environment.

For instance, a data scientist might be exceptionally skilled at exploratory data analysis and visualization but find model deployment and MLOps challenging. The manager can leverage their strengths by assigning them to projects where those skills are paramount, while providing structured support or pairing them with colleagues who have expertise in deployment for the aspects they find difficult. This approach not only helps the individual grow but also optimizes the team's overall productivity. The feedback loop here is vital; the individual needs to understand not only what they are doing well, but also precisely where and how they can improve, with clear, actionable steps.

In AI-integrated teams, where projects often involve complex interdependencies and rapid change, fostering accountability is paramount. This means ensuring that individuals understand their responsibilities and the impact of their work on the broader team and project objectives. When tasks are delegated, especially those involving AI components, leaders must ensure clarity on ownership and deliverables.

Example: In a fraud detection project, responsibilities must be clearly delineated:

- **Member A:** Data ingestion and cleaning.
- **Member B:** Feature engineering and selection.
- **Member C:** Model training and validation.
- **Member D:** Deployment and real-time monitoring.

Each individual must understand how their work feeds into the next stage. Regular check-ins are critical for maintaining accountability. These are not just opportunities to track progress, but also to address any blockers or challenges that might be hindering an individual's ability to meet their commitments. If a team member is consistently missing deadlines or delivering work that requires significant rework, the manager must address this directly and collaboratively.

This structured approach to performance management, grounded in clear expectations, measurable outcomes, and consistent, actionable feedback, is what differentiates a functional AI team from a truly high-performing one. It creates a transparent and supportive environment where individuals feel empowered to contribute their best. The integration of AI in tracking and analysis can enhance this process by providing objective data, but the human element—empathy, judgment, and the ability to foster growth—remains irreplaceable.

Ethical Leadership in AI Team Dynamics

The integration of artificial intelligence into team dynamics introduces a complex web of ethical considerations. As we transition to fostering an environment of responsible AI utilization, the emphasis shifts to ensuring that tools designed to augment human capabilities do not inadvertently undermine fairness, equity, or trust. Ethical leadership acts as the bedrock upon which high-performing, yet morally sound, AI-integrated teams are built.

One of the most immediate ethical challenges arises when AI tools are employed in performance evaluation or management processes.

While AI can offer objective data points, its application must be scrutinized to prevent unintended biases. For instance, an AI system designed to analyze productivity might inadvertently penalize individuals whose work patterns differ from the norm—perhaps those who take more time for deep thinking or collaborative problem-solving, which are crucial in complex AI development.

Such systems can also be trained on historical data that reflects existing societal biases, leading to discriminatory outcomes. A leader's ethical responsibility here is to ensure that any AI used in performance assessment is rigorously audited for fairness and transparency. This involves:

- **Understanding the algorithms:** Knowing how the tool reaches its conclusions.
- **Auditing the data:** Ensuring training data does not reflect historical prejudice.
- **Maintaining human oversight:** Ensuring the final decision always rests with a human manager who can interpret qualitative contributions that AI might miss.

Leaders must champion the use of AI as a supplementary tool for human judgment. Fostering an environment where team members feel that their evaluations are fair and unbiased, even when AI is involved, is a critical component of building trust. This requires open communication about how AI is being used, what data is being collected, and how that data informs decisions.

Beyond individual performance, ethical leadership extends to how AI influences team composition and task allocation. The temptation to use AI for optimizing team assignments—perhaps by predicting who would be most efficient on a particular task or by dynamically reassigning members based on perceived skill gaps—can be intense. However, this practice must be approached with extreme caution. Algorithmic bias can creep into these assignments, leading to certain team members being consistently overlooked for challenging or high-visibility projects, thereby hindering their professional development.

This could manifest as an AI system favoring individuals whose past performance, as recorded in training data, aligns with a narrow definition of success, inadvertently excluding those with creative approaches or diverse skill sets. An ethical leader must ensure that AI-driven team assignments are transparent and that they actively promote diversity and inclusion. Instead of letting an algorithm make these decisions unilaterally, leaders should use AI insights as one data point among many, alongside their own understanding of team members' aspirations, growth potential, and interpersonal dynamics.

The goal should be to use AI to identify potential synergies or development opportunities, rather than to automate assignment decisions that have significant career implications. Maintaining fairness in task distribution is not just about equity; it is about maximizing the team's collective intelligence and fostering an environment where everyone has the opportunity to contribute and grow.

Transparency regarding the role of AI within the team is another cornerstone of ethical leadership. Team members have a right to know when and how AI is being used, especially when it impacts their work, their development, or decisions made about them. This includes understanding which AI tools are used for tasks such as code review, project management, data analysis, and even communication pattern analysis, and what information these tools process.

A lack of transparency can breed suspicion and distrust, leading to a reluctance to adopt new technologies or a perception that decisions are being made by an opaque "black box." Leaders must proactively communicate the purpose and limitations of AI tools, explaining their benefits while acknowledging potential drawbacks and areas where human oversight remains crucial. For instance, if an AI tool flags potential issues in code, the team should be informed that this is a suggestion and that a human code reviewer will make the final determination. This transparency builds a foundation of trust, enabling team members to engage with AI technologies more constructively and to provide valuable feedback on their effectiveness and ethical implications. It transforms AI from a potential source of anxiety into a collaborative partner, deliberately understood and utilized.

Fostering an environment of integrity and trust requires leaders to actively engage their teams in discussions about the ethical implications of their AI work. This is not a passive endeavor; it involves creating safe spaces for dialogue where concerns can be raised without fear of reprisal. For teams developing AI systems, this means encouraging critical thinking about the potential societal impacts of their creations, including privacy and security issues, as well as the risk of misuse or unintended consequences.

Leaders should facilitate workshops, provide resources, and encourage open debate on ethical dilemmas. This engagement empowers team members to become ethical stewards of the technology they are building, rather than simply implementers. When individuals understand and are comfortable with the technologies they are using, and when they feel their ethical concerns are heard and valued, they are more likely to be engaged and committed to their work. This proactive approach to ethics integration embeds it into the team's culture, ensuring that responsible innovation is not an afterthought but a core principle guiding all AI-related activities.

The "Gravity of Command" in the context of AI-integrated teams emphasizes that leaders bear the ultimate responsibility for ensuring that technological advancements are pursued with a deep consideration for human well-being and fairness. This means that leaders must not only understand the technical aspects of AI but also its ethical dimensions and societal implications. They must set a clear tone from the top, demonstrating their commitment to ethical AI practices through their own actions and decisions. This involves making difficult choices, such as opting for a less efficient but more ethical AI solution or investing in additional human oversight to mitigate AI-driven risks. It also means holding themselves and their teams accountable for any ethical missteps. By championing responsible AI development and deployment, leaders can ensure that technological progress genuinely enhances human potential and societal good, rather than creating new forms of inequality or harm.

Ultimately, leading AI-integrated teams ethically is about maintaining a human-centric approach in an increasingly automated world. It requires leaders to be vigilant, inquisitive, and deeply committed to fairness. This involves:

- **Continuous Education:** Keeping pace with the evolving landscape of AI ethics.
- **Bias Hunting:** Actively seeking out potential biases in data and outcomes.
- **Promoting Transparency:** Ensuring the "why" behind AI-assisted decisions is clear.
- **Lifecycle Integration:** Embedding ethical checkpoints from data collection to deployment.

When leaders prioritize integrity, trust, and human well-being, they not only build high-performing teams but also ensure that the powerful tools of artificial intelligence are harnessed for the betterment of all. This unwavering focus on ethical leadership is what truly differentiates an ordinary AI team from one that is genuinely exceptional and socially responsible.

The Talent & Culture Checklist

- [] **The Job Description Audit:** Review your open roles. Are you asking for 10 years of experience in a tool that is only two years old? **Action:** Update requirements to prioritize "proven adaptability" and "first-principles thinking" over stagnant legacy metrics.
- [] **The "Drudgery Audit":** Ask your team: "What is the one repetitive task you hate the most?" **Action:** Task them with finding an AI tool to automate that specific task this week. Shift the mindset from "working harder" to "working smarter."

- [] **The "Co-Pilot" Policy:** Explicitly authorize your team to use AI tools for drafting, coding, and brainstorming. **Action:** Formalize a policy that removes the fear that "using AI is cheating" and replaces it with the mandate that "leveraging AI is essential."

*****

Summary

Earth's Grounding Command is about building a team that is rooted enough to withstand the pressure of high-speed delivery, but flexible enough to grow into new shapes.

Bob hired the candidate with a philosophy degree. Six months later, she found a logic flaw in the AI's reasoning that the senior engineers had missed. She did not know the legacy stack, but she knew how to think. By prioritizing the "spark" of curiosity over the "history" of a resume, Bob didn't just fill a seat; he future-proofed his team.

In the next chapter, **Moonlight Innovation**, we will look at how to use this high-performing team to see in the dark—using AI for strategic decision-making and navigating the unknown.

5. Moonlight Innovation: Strategic Decision-Making with AI Insights

Seeing in the Dark

"The data doesn't make sense, Alice. We've sold this product for ten years. Our customers want stability, not a generative design tool."

Bob tapped the printed quarterly report. The sales figures were solid, and the customer feedback was consistent. To the naked eye, the path forward was clear: stay the course.

Alice pointed to the screen, where a complex web of nodes and heat maps pulsed. "That's the daylight view, Bob. You're looking at what happened yesterday. But look at the moonlight view."

She zoomed in. "The AI has been scanning patent filings, GitHub repositories, and niche developer forums. It's picking up a weak signal: a 400% spike in discussions about 'autonomous workflows' in the last three months. It's not in our sales data yet because these aren't our customers... yet. But they will be in the market in two years. If we wait for the daylight data, we'll be obsolete."

Bob leaned in. The "safe" path suddenly looked dangerous.

This is Moonlight Innovation. In the business world, "Daylight" is the data you already have: your P&L, your backlog, and your current customers. It is bright, clear, and comfortable. "Moonlight" is the realm of AI. It illuminates the shadows, including emerging trends, hidden risks, and weak signals that suggest a storm is coming long before it hits the radar.

In the Gravity of Command, strategic leadership is about using AI to see in the dark without losing your footing in the real world.

The Leadership Evolution: From Oveisight to Orchrestation

From Intuition to Insight: The New Strategic Radar

Historically, strategy was often a battle of intuitions. The loudest voice or the highest-paid person's opinion (HiPPO) won the debate. Today, AI transforms strategy from a guessing game into a discipline of predictive probability.

1. Forecasting the Unseen: Traditional forecasting looks at historical sales and draws a straight line forward. AI forecasting looks sideways. By ingesting vast external datasets, such as news articles, geopolitical risk reports, weather patterns, and competitor job postings, AI can predict market shifts that have nothing to do with your past performance.

- **The Capability:** An AI system can detect a convergence of breakthroughs in materials science and AI that might disrupt your supply chain two years from now, giving you the lead time to pivot.

2. The R&D Compass: Deciding where to invest limited capital is the "Tetris" of leadership. Should you fund Project A or Project B? AI acts as an unbiased arbiter. It can analyze the "success probability" of an R&D initiative by comparing it against thousands of historical projects, patent landscapes, and market saturation levels.

- **The Result:** You stop funding "pet projects" and start funding "probable wins."

The Black Box Challenge: When to Trust the Machine

However, Moonlight can be tricky. Shadows can look like monsters. Just because the AI predicts a trend does not mean you should bet the company on it. This brings us back to the Gravity of Command: You must balance Algorithmic Insight with Human Judgment.

The "Why" Test: AI is fantastic at correlation (people who buy X also buy Y), but terrible at causation (why they do it).

- **Scenario:** The AI recommends reallocating 50% of the marketing budget to a new digital channel because the "conversion metrics" look high.
- **The Human Check:** Bob, drawing on his experience, recognizes that the high metrics stem from a "click-bait" strategy that damages the brand's long-term reputation. The AI sees the click; the leader considers the trust eroding.

Leadership Rule: Use AI to generate the options, but use your judgment to make the choice. If the AI cannot explain why a strategy is sound (the "Black Box" problem), do not follow it unthinkingly.

*****

[Sidebar] Gravity Pulls: The Strategic Filter

- **Data over HiPPO:** Use AI to challenge the "Highest Paid Person's Opinion." When you bring data to a gunfight, you change the dynamic.
- **The Horizon Scanner:** Don't just use AI to optimize today (efficiency); use it to scan tomorrow (opportunity). If your AI is only cutting costs, you are missing half the value.

- **The Ethical Brake:** AI will continually optimize for the metric you give it, such as profit. It is your job to apply the "Ethical Brake" to ensure the strategy aligns with your values.

*****

Building the Business Case: Selling the Moon

Identifying the innovation is only step one. Step two is getting the budget. Alice knew the data was correct, but she had to convince the CFO. This requires the art of the AI Business Case.

Many leaders fail here because they focus on the "cool factor" of the tech. Stakeholders do not care about the algorithm; they care about the ROI (Return on Investment).

How to Frame AI Investment:

1. **Quantify the Invisible:** Don't just say "better decisions." Say, "By using AI to predict supply chain disruptions, we can reduce inventory holding costs by 15%."
2. **The Cost of Inaction:** Use the AI's data to show what happens if you don't move. "Competitor X has already filed patents in this space. If we wait 12 months, the cost of entry triples."
3. The Competitive Moat: Explain how this AI initiative builds a long-term advantage that competitors cannot easily copy, such as a proprietary data set.

Summary

Alice's weak signal turned out to be right. Six months later, the "autonomous workflows" trend took off, and because Bob had green-lit a small pilot, the company already had a prototype in market. Moonlight Innovation does not replace strategy with software; it gives leaders a telescope so they can see further than instinct alone.

AI extends this telescope into forecasting and risk management by absorbing vast, real-time data—from markets and supply chains to news sentiment—to spot threats and opportunities earlier. It puts powerful analytics into everyday tools so managers can ask plain-language questions, get clear answers, and act faster. Used well, AI improves accuracy, speeds up decisions, reveals options humans might miss, and helps leaders direct resources where they matter most.

But these advantages only hold if the data is clean, the models are monitored, and human judgment stays in the loop. Leaders must demand explainable systems, build data-literate teams, and treat ethics and bias checks as non-negotiable. When those foundations are in place, AI stops being a black box and becomes a strategic partner—turning weak signals into early moves instead of missed chances.

*****

Evaluating the AI Landscape: A Strategic Framework

The imperative to judiciously integrate artificial intelligence into organizational strategy requires a rigorous, systematic approach to evaluating AI tools and technologies. This section aims to equip leaders with a comprehensive framework for these assessments, ensuring that every technological adoption is a deliberate, strategic move aligned with overarching business objectives rather than a mere response to prevailing trends. Moving beyond the theoretical benefits of AI, we now pivot to the practicalities of selection, focusing on how to discern which AI solutions offer genuine strategic fit and return on investment.

At the core of this evaluation lies a nuanced understanding of the diverse landscape of AI technologies. Managers must differentiate among AI categories, recognizing their distinct capabilities and limitations. Broadly, we can consider supervised, unsupervised, and reinforcement learning.

- **Supervised Learning:** This excels in tasks where historical data is abundant and labeled, such as image recognition, spam filtering, or predictive analytics for well-defined outcomes. If an organization seeks to automate customer service by classifying incoming queries or predicting sales based on past performance, supervised learning models would be a prime candidate.

- **Unsupervised Learning:** Conversely, this is powerful for discovering hidden patterns and structures in unlabeled data, making it ideal for customer segmentation, anomaly detection, or market basket analysis. For a business looking to understand distinct customer groups without pre-defined labels or to identify fraudulent transactions by spotting deviations from normal behavior, unsupervised learning offers unparalleled insights.
- **Reinforcement Learning:** Often employed in dynamic environments, this focuses on learning through trial and error, optimizing decision-making sequences to achieve a specific goal. This is particularly relevant for areas such as algorithmic trading, robotic control, and dynamic pricing strategies, where continuous adaptation is key. A retail company looking to optimize its promotional pricing in real-time based on competitor actions and inventory levels might explore reinforcement learning solutions.

Beyond these foundational machine learning paradigms, the AI ecosystem encompasses other critical technologies, such as Natural Language Processing (NLP), Computer Vision, and Expert Systems.

- **Natural Language Processing (NLP):** This enables machines to understand, interpret, and generate human language, unlocking opportunities in areas such as sentiment analysis of customer reviews, automated document summarization, or conversational AI for chatbots. For an organization aiming to gauge public perception of its brand or to streamline internal knowledge management, NLP-powered tools are indispensable.
- **Computer Vision:** This empowers machines to "see" and interpret visual information, with applications in quality control on manufacturing lines, object detection for autonomous vehicles, and analysis of medical imagery. A

manufacturing firm seeking to enhance product defect detection with greater speed and accuracy than human inspectors would benefit from computer vision solutions.

- **Expert Systems:** While an older form of AI, these still hold relevance for encoding domain-specific knowledge and automating complex decision-making processes in fields like medical diagnosis or financial advisory services, though they are often now augmented by modern machine learning techniques.

Once the conceptual landscape of AI technologies is understood, the next crucial step is to assess their potential Return on Investment (ROI). This is not merely about quantifying cost savings or revenue increases, though these are vital; it also involves considering the strategic value an AI solution brings, such as enhanced customer experience, improved operational efficiency, accelerated innovation, or a more decisive competitive advantage.

A framework for ROI calculation should encompass direct financial benefits, such as reduced labor costs or increased sales; indirect benefits, such as improved employee satisfaction or better decision-making quality; and intangible benefits, such as enhanced brand reputation or first-mover advantage. It is also critical to factor in the total cost of ownership, which extends beyond the initial purchase price to include implementation, integration, training, maintenance, and potential upgrades to the data infrastructure.

A solution that promises significant gains but requires an exorbitant investment in new hardware and specialized personnel might not present a favorable ROI in the short to medium term. Leaders must cultivate a disciplined approach to ROI assessment, acknowledging that not all AI investments will yield immediate or easily quantifiable financial returns, especially in areas focused on long-term research and development or foundational capability building. The strategic value might manifest over several years, requiring a patient, forward-looking perspective.

When evaluating vendor solutions, a systematic approach is essential. Managers should develop a robust set of criteria that goes beyond feature checklists. Key considerations include:

1. Problem-Solution Alignment: Does the AI tool directly address a well-defined business problem or strategic objective? Is there a clear and demonstrable link between the AI's capabilities and the desired outcome? This requires a deep understanding of the business need before even looking at potential solutions. For instance, if the goal is to reduce customer churn, an AI solution that excels at predicting customer behavior based on usage patterns and interaction history would be more relevant than one focused on image generation.

2. Data Requirements and Readiness: AI systems are data-hungry. The vendor's solution will have specific data input requirements. Must organizations assess their current data landscape: Is the necessary data available? Is it of sufficient quality, completeness, and accuracy? Are there privacy or regulatory constraints on data usage? A solution requiring extensive data cleansing or the acquisition of new data sources might significantly increase implementation time and cost, impacting ROI. Vendors should ideally provide clear guidance on data preparation and integration.

3. Technical Capabilities and Performance: This involves scrutinizing the underlying AI models, algorithms, and architecture. How accurate is the AI in its predictions or classifications? What is its performance under varying loads or conditions? Is the AI explainable? For critical applications, the ability to understand *why* an AI makes a particular recommendation (explainable AI or XAI) is paramount for building trust and ensuring accountability. Benchmarking performance against industry standards or internal pilot projects is advisable.

4. Scalability and Flexibility: Will the solution be able to grow with the organization's needs? Can it handle increasing volumes of data and users? Is it flexible enough to adapt to evolving business requirements and integrate with other systems? A solution that is easy to scale up (or down) without prohibitive costs or significant architectural changes offers greater long-term value. Cloud-native AI solutions often offer inherent scalability advantages.

5. Integration Complexity and Compatibility: How easily can the AI tool be integrated into existing IT infrastructure and workflows? Does it require extensive custom development or compatibility with proprietary systems? The less disruptive the integration, the faster the time-to-value and the lower the implementation risk. Open APIs and standard integration protocols are generally indicators of better integration potential. Compatibility with existing data lakes, CRM systems, ERP platforms, and BI tools is crucial.

6. Vendor Viability and Support: Beyond the technology itself, the vendor's reputation, financial stability, track record, and the quality of their customer support and training are critical. Will the vendor be around to support the solution in the long term? What is their roadmap for future development? Is their support responsive and knowledgeable? Case studies and client references can provide valuable insights into the vendor's reliability.

7. Total Cost of Ownership (TCO) and ROI Justification: As mentioned earlier, a comprehensive assessment of TCO is vital. This includes not only license or subscription fees but also implementation services, infrastructure costs (hardware and cloud services), data management, training, ongoing maintenance, and support. The vendor should be able to clearly articulate the expected ROI and the assumptions underpinning their projections. Managers must conduct their own rigorous ROI analysis to validate the vendor's claims.

8. Ethical and Security Considerations: Does the AI solution comply with relevant data privacy regulations (e.g., GDPR, CCPA)? Are there robust security measures in place to protect sensitive data processed by the AI? Critically, are there mechanisms to identify and mitigate potential biases within the AI's algorithms and outputs? For instance, an AI used for hiring decisions must be rigorously checked for biases that could discriminate against certain demographic groups. Vendors should be transparent about their approach to AI ethics and security.

Evaluating these criteria systematically allows for a more informed and strategic selection process. This often involves a multi-stage approach, starting with a broad market scan, followed by shortlisting potential vendors, conducting proof-of-concept (POC) projects or pilot implementations, and finally making a well-justified selection. A POC allows an organization to test a specific AI tool's capabilities on a real-world business problem with its own data, providing invaluable insights into its performance, integration feasibility, and actual value proposition before committing to a large-scale deployment.

This pragmatic approach mitigates risk and ensures that AI investments are purposeful, contributing directly to achieving long-term strategic objectives rather than being adopted merely for the sake of technological novelty. Prudent leadership demands that such technology investments are anchored in a clear understanding of their strategic implications and their potential to drive measurable business outcomes.

Moreover, the evaluation process must consider the organizational readiness for AI adoption. This includes assessing the existing talent pool, the culture's receptiveness to data-driven decision-making, and the availability of necessary skills for implementation and ongoing management. If the organization lacks in-house expertise in data science, AI engineering, or ethical AI governance, the evaluation should also encompass the cost and feasibility of acquiring or developing such capabilities. This might involve hiring new talent, upskilling existing employees, or partnering with external consultants. A sophisticated AI tool, however powerful, will underperform if the organization lacks the human capital to deploy, manage, and interpret its outputs effectively. Therefore, the evaluation of AI tools should be intertwined with an assessment of organizational capacity and a clear plan for talent development and cultural adaptation.

The strategic fit of an AI technology is not a static determination. As the business environment evolves, new data sources become available, and AI technology itself advances, periodic re-evaluation of existing AI investments becomes necessary. Technologies that were once at the cutting edge may become legacy systems, or new business priorities may supersede their original strategic purpose.

This continuous assessment ensures that AI investments remain aligned with the organization's dynamic strategic landscape, preventing technological obsolescence and ensuring AI continues to drive competitive advantage rather than become a costly burden. Such ongoing scrutiny reinforces the idea that AI adoption is not a one-time event but a continuous journey of adaptation and optimization, demanding a leadership mindset that is both visionary and pragmatic. It is about ensuring that technology remains a powerful enabler of strategy, not an end in itself.

In essence, evaluating AI tools and technologies for strategic fit is a multidimensional undertaking. It requires deep domain knowledge, a clear understanding of AI capabilities and limitations, a rigorous approach to ROI and TCO analysis, meticulous vendor vetting, and a keen awareness of organizational readiness and ethical imperatives. By systematically applying these principles, leaders can ensure that their investments in artificial intelligence are not speculative bets but calculated, strategic moves that propel the organization toward its long-term objectives, solidifying its position in an increasingly complex and competitive global marketplace. This disciplined approach to technology selection is a hallmark of effective strategic leadership in the age of AI.

Forecasting Future Trends and Market Dynamics

The business landscape is in perpetual motion, shaped by an ever-accelerating pace of technological advancement and evolving market demands. For organizations aiming not just to adapt but to lead, the ability to anticipate these shifts is paramount. This is where the predictive power of Artificial Intelligence moves from the realm of operational efficiency and customer engagement to a more profound strategic function: the forecasting of future trends and market dynamics.

AI, with its capacity to process and discern patterns within colossal and complex datasets far beyond human cognitive limits, offers an unprecedented lens through which to view the horizon of possibility and potential disruption. It transforms foresight from an art relying on intuition and limited historical analysis into a data-driven science, enabling strategic decision-makers to move from reactive maneuvering to proactive shaping of their organization's future.

The foundation of AI-driven trend forecasting lies in its ability to ingest and analyze a vast spectrum of data sources. Consider the sheer volume of information generated daily:

- **Academic research papers** detailing nascent technological breakthroughs.
- **Patent filings** hint at future product innovations.
- **News articles** reporting on geopolitical shifts.
- **Consumer reviews and social media sentiment** provide real-time feedback.
- **Granular supply chain data** and competitor strategic announcements.

Historically, a human analyst might painstakingly review a fraction of this information, relying on heuristics and personal networks to identify emerging patterns. AI, however, can ingest, categorize, and analyze these diverse data streams simultaneously. Machine learning algorithms, particularly those focused on natural language processing (NLP) and time-series analysis, can identify subtle correlations and leading indicators that might otherwise go unnoticed.

For instance, an NLP model could monitor scientific publications and patent databases for keywords related to material science or biotechnology. By tracking the frequency, co-occurrence, and sentiment associated with these terms, along with the geographic and institutional origins of these filings, an AI system can begin to map the emergence of novel scientific fields and predict which areas are likely to yield commercially viable technologies in the coming years. This granular analysis can flag the early-stage development of a new battery technology, a novel drug-delivery mechanism, or a breakthrough in quantum computing, providing significant lead time for strategic investment or partnership.

Furthermore, AI can extend its predictive reach into market dynamics by analyzing consumer behavior and economic indicators with a sophistication previously unimaginable. By examining online purchasing patterns, search queries, social media interactions, and even aggregated foot traffic data (where ethically sourced and anonymized), AI can identify shifts in consumer preferences, emerging lifestyle trends, and unmet needs.

For example, an AI model analyzing e-commerce data might detect growing interest in sustainable fashion, not just by tracking direct searches for "eco-friendly clothing," but by identifying subtle shifts in purchases of related items such as organic cotton, recycled materials, or brands that emphasize ethical sourcing. It can also correlate these shifts with demographic information and geographic location to pinpoint specific market segments driving these trends.

Beyond consumer behavior, AI can integrate macroeconomic data, geopolitical risk assessments, and supply chain resilience metrics to forecast broader market shifts. An AI system might predict a potential bottleneck in the global semiconductor supply chain due to rising geopolitical tensions in a key manufacturing region, combined with an anticipated surge in demand from the burgeoning electric vehicle market. This foresight allows organizations to proactively secure supply lines, diversify manufacturing bases, or even accelerate the development of alternative technologies.

The integration of these AI-driven foresight capabilities into strategic planning processes is not merely about receiving a report; it requires a fundamental shift in how organizations approach decision-making. Instead of relying solely on historical performance and current market snapshots, strategic planning sessions can be augmented by AI-generated scenarios and trend forecasts.

Imagine a quarterly strategic review where the leadership team is presented not only with past sales figures but also with AI-identified emerging consumer needs in a related product category, projected shifts in regulatory landscapes affecting raw material sourcing, and a competitive analysis that flags a smaller, agile competitor making significant inroads in a niche market using a novel technology. This provides a richer, more dynamic context for decision-making.

The AI can present multiple potential future scenarios, each with associated probabilities and possible impacts on the business. For instance, an AI might generate a scenario in which widespread adoption of augmented reality glasses leads to a significant decline in demand for traditional smartphones within 5 to 7 years, alongside a counter-scenario in which the primary interface shifts to voice-activated smart home devices. Leaders can then use these AI-generated insights to stress-test their current strategies, identify potential vulnerabilities, and proactively explore new avenues for growth.

To effectively leverage AI for trend forecasting, organizations must cultivate a specific set of capabilities:

1. **Data Infrastructure and Governance:** This involves establishing robust data pipelines that can ingest, clean, and store diverse datasets from internal and external sources. Ensuring data quality, accuracy, and adherence to privacy regulations is non-negotiable. Without high-quality, reliable data, AI models will produce flawed or misleading insights.
2. **Advanced Analytics and AI Expertise:** This means having access to data scientists, AI engineers, and domain experts

who can build, train, and interpret complex forecasting models. It is not just about deploying off-the-shelf tools; it is about tailoring AI solutions to specific business contexts and industry nuances.

3. **A Culture of Experimentation and Data Literacy:** Leaders and employees at all levels must be comfortable with data-driven insights, willing to challenge assumptions, and open to acting on probabilistic forecasts. This often involves training and fostering a mindset that embraces continuous learning and adaptation.

Consider the application in the automotive industry. An AI system could analyze trends in battery technology research, government subsidies for electric vehicles, consumer preferences for vehicle range and charging speed, and patent filings for new autonomous-driving software. By correlating these data points, the AI might predict a significant shift toward solid-state batteries over the next 8 years, leading to a dramatic increase in vehicle range and a decrease in charging times. This foresight would enable an automotive manufacturer to strategically invest in solid-state battery research and development, secure partnerships with emerging battery technology companies, or even begin redesigning vehicle platforms to accommodate these new battery architectures.

Simultaneously, the AI could highlight a rising consumer demand for personalized in-car digital experiences, driven by social media trends and the increasing integration of personal devices into daily life. This would prompt the manufacturer to prioritize investment in advanced infotainment systems, seamless connectivity, and AI-powered personal assistants within vehicles, ensuring they remain competitive in a rapidly evolving market.

Another compelling example can be found in the pharmaceutical sector. AI can scan global scientific literature, clinical trial databases, and genomic research to identify emerging therapeutic targets for diseases that are currently underserved. By analyzing research papers on novel protein interactions or genetic markers associated with rare diseases, combined with market analysis of unmet medical needs and the economic viability of developing new treatments, AI can flag promising areas for drug discovery.

For instance, an AI might identify a cluster of research papers exploring the efficacy of certain repurposed compounds for neurodegenerative diseases, alongside an increasing incidence of these diseases in aging populations and a lack of effective treatments. This foresight could guide a pharmaceutical company's R&D pipeline, directing resources toward developing new drug candidates for these conditions years before they become mainstream therapeutic targets. Furthermore, AI can analyze anonymized, aggregated patient data to predict shifts in healthcare delivery models, such as a move toward more personalized medicine or increased demand for remote patient monitoring, enabling companies to develop relevant diagnostic tools, digital health platforms, or specialized treatment plans.

The proactive adoption of AI-driven forecasting allows organizations to achieve a decisive competitive advantage. Instead of reacting to market disruptions, they can anticipate them and position themselves to either lead the transition or mitigate its negative impacts. This might involve developing new product lines, entering emerging markets before competitors, divesting from assets that are likely to become obsolete, or forging strategic alliances to acquire necessary capabilities.

For instance, a retail company might use AI to forecast a shift from brick-and-mortar stores to online and experiential retail, driven by changing consumer habits and the rise of augmented reality shopping. Armed with this insight, the company could proactively invest in its e-commerce platform, develop immersive online shopping experiences, and reimagine its physical stores as experiential hubs rather than mere points of sale.

This proactive approach not only safeguards the organization against future uncertainties but also unlocks new avenues for innovation and growth, ensuring its long-term relevance and success in an unpredictable global marketplace. The strategic value of AI in forecasting future trends and market dynamics cannot be overstated; it is rapidly becoming an indispensable tool for navigating the complexities of the modern business environment and securing a position at the vanguard of innovation. It empowers leaders to see not just what is happening now, but what is likely to happen next, and to act with confidence and foresight.

The modern enterprise operates under the constant imperative to do more with less, striving to achieve greater impact with constrained resources. This fundamental challenge of resource allocation—whether it pertains to financial capital, human talent, technological infrastructure, or time—is a perennial strategic concern. Historically, organizations have relied on a combination of experience, intuition, and often rigid, backward-looking analytical methods to make these critical decisions. However, the increasing complexity and velocity of business environments demand a more sophisticated, dynamic, and precise approach. This is precisely where Artificial Intelligence emerges as a transformative force, moving beyond mere automation to become an intelligent partner in optimizing the deployment of an organization's most valuable assets. AI-driven resource allocation involves intelligent, predictive, and adaptive strategies that align every available resource with initiatives that promise the greatest strategic impact and return on investment.

At the core of AI's capability in resource optimization lies its unparalleled ability to process and analyze vast, multi-dimensional datasets to uncover intricate relationships and predict future outcomes. Consider the challenge of capacity planning. Traditionally, this might involve extrapolating past performance, applying industry benchmarks, and making educated guesses about future demand. AI, however, can ingest a far richer tapestry of data, including historical project completion rates, team member skill sets, task complexity, and project dependencies. By analyzing these factors, AI-powered models can develop highly accurate forecasts of resource needs. For instance, an AI system tasked with capacity planning for a software development department could analyze the estimated effort for each feature request while cross-referencing the specialized skills required for each task. The AI can then predict not only the total hours needed but also the specific skill sets that will be in highest demand and when those peaks will occur. This allows management to proactively address potential bottlenecks, such as identifying a shortage of experienced engineers, and to initiate recruitment or training programs well in advance to prevent project delays and team burnout.

Furthermore, AI excels at the intricate task of workload balancing. In any organization, there is a natural tendency for certain individuals or teams to become overloaded while others have spare capacity, an imbalance that reduces efficiency and stifles morale. AI addresses this by dynamically analyzing real-time workloads across teams and individuals, factoring in the complexity, priority, and estimated time to completion for each assigned task. For a marketing department, an AI could monitor the progress of various campaign initiatives and the demand for creative assets. If the system detects that the graphic design team is consistently facing delays due to an overwhelming number of requests, it can recommend reallocating some of these tasks to other teams with available capacity or suggest prioritizing requests based on their projected impact on marketing objectives. More sophisticated systems can even optimize task sequencing to minimize idle time and context-switching. This intelligent distribution of work ensures that human capital is utilized effectively, preventing talent from being squandered on low-impact activities.

A compelling application of AI in resource allocation is in the realm of investment prioritization. Organizations constantly face decisions about where to invest their limited capital, such as which research and development projects to fund or which infrastructure upgrades to undertake. AI provides a data-driven framework for high-stakes decision-making by quantifying the potential return on investment for various initiatives. This involves building predictive models that incorporate a wide array of variables. For a technology company, an AI might analyze the potential market size for a new product feature, projected development costs, and the likelihood of successful adoption using historical data. If a company is considering two different product enhancements, the AI could analyze customer feedback and technical complexity for each. Based on this analysis, the AI might project that a user interface enhancement has a higher potential to drive engagement than a technical speed enhancement, despite the latter being a more significant engineering feat. This analytical rigor helps shift investment decisions away from subjective preferences toward objective, data-backed strategic choices, ensuring that capital is deployed where it can yield the greatest benefits.

Moreover, AI's ability to adapt and learn means that these resource allocation strategies are not static. As new data becomes available—whether it's updated project progress reports, shifts in market sentiment, or emerging technological capabilities—AI models can continuously re-evaluate and adjust recommendations. This dynamic reallocation is crucial in fast-paced industries. Imagine a scenario in which a company allocates a significant portion of its marketing budget to a particular advertising channel based on initial projections. However, as the campaign progresses, an AI monitoring real-time campaign performance, customer response rates, and competitor activities might detect a significant shift in consumer engagement towards a different, previously less prioritized channel. The AI could then recommend an immediate reallocation of budget from the underperforming channel to the one showing superior results, maximizing the campaign's overall effectiveness and preventing wasted expenditure. This agility allows organizations to pivot quickly in response to changing circumstances, ensuring that resources are always directed towards the most promising opportunities.

Implementing AI for resource optimization also requires careful consideration of the underlying data infrastructure and the ethical implications of algorithmic decision-making. Robust data pipelines are essential to feed AI models with accurate, timely, and comprehensive information. This includes integrating data from various enterprise systems, such as ERP, CRM, HRIS, and project management tools, as well as external sources. Data governance policies must be in place to ensure data quality, privacy, and security. Furthermore, as AI takes on more decision-making roles in resource allocation, it is crucial to ensure transparency and fairness. Algorithmic bias can inadvertently lead to the under-resourcing of specific teams or projects based on historical data that reflects past inequities. Therefore, organizations must actively audit their AI models for bias and establish human oversight mechanisms to review and, if necessary, override AI-driven recommendations, especially in sensitive areas such as personnel allocation. The goal is not to replace human judgment entirely but to augment it with powerful, data-driven insights, creating a collaborative process that leverages the strengths of both AI and human intelligence.

Consider the intricate process of managing a large-scale construction project. This involves allocating billions of dollars in capital, coordinating thousands of workers with diverse skill sets, and managing vast quantities of materials and heavy equipment. Traditionally, this has been a mammoth undertaking reliant on experienced project managers, Gantt charts, and manual resource tracking. AI can revolutionize this. By integrating real-time data from IoT sensors on construction sites, monitoring equipment usage, material delivery, and progress, along with weather forecasts and supply chain updates, AI can build a dynamic model of the entire project. The system can predict the precise timing and amount of concrete needed for specific structural elements, accounting for weather conditions that might delay pouring. It can identify if a shortage of skilled electricians is imminent and recommend cross-training or expedited hiring. It can optimize the deployment of heavy machinery, ensuring that cranes and excavators are used efficiently across different sections of the site rather than sitting idle. Furthermore, AI can analyze the critical path, identify the tasks most susceptible to delay, and proactively suggest resource reallocations to mitigate risks. For instance, if a shipment of specialized steel beams is delayed, the AI could analyze alternative construction sequences to keep the overall timeline on track, thereby preventing costly overruns.

In the healthcare sector, AI-driven resource allocation can be life-saving. Hospitals and healthcare systems face immense pressure to manage limited resources—beds, specialized equipment, surgical teams, and pharmaceuticals—while simultaneously providing high-quality patient care. AI can help optimize patient flow and resource utilization by analyzing real-time data on patient admissions, acuity levels, discharge rates, and staff availability. For example, an AI system could forecast an impending surge in demand for ICU beds based on an uptick in respiratory illnesses in the community, allowing administrators to proactively prepare by identifying surge capacity options. In the operating room, AI can help schedule surgeries more efficiently by considering the availability of surgical teams with specific expertise, required equipment, and the estimated duration of procedures, while also factoring in the urgency of each case. This not only maximizes the use of expensive surgical suites and skilled personnel but also reduces patient wait times. Furthermore, AI can optimize the allocation of critical medical supplies by predicting demand for pharmaceuticals and personal protective equipment based on patient demographics and seasonal trends, thereby preventing stockouts and ensuring that essential resources are available when and where they are needed most.

The strategic advantage gained from optimizing resource allocation with AI extends to every facet of an organization. It allows for more agile responses to market shifts, more efficient execution of projects, and ultimately, a higher likelihood of achieving strategic objectives. By moving from reactive, often intuitive resource decisions to proactive, data-driven, and continuously learning optimization strategies, businesses can unlock new levels of productivity, innovation, and competitive strength. This intelligent deployment of resources is not merely an operational improvement; it is a fundamental enhancement of an organization's strategic command and control capabilities, enabling it to navigate complexity and achieve its goals with unprecedented precision and effectiveness. The ability to intelligently allocate budget, personnel, and infrastructure ensures that every element of the organization works in concert, directed toward the most impactful outcomes, thereby transforming potential into tangible success.

The strategic integration of Artificial Intelligence into decision-making processes, while immensely powerful, does not mandate the abandonment of human oversight. Instead, it signifies a profound evolution in how judgment is exercised. The insights provided by AI are, by their very nature, derived from patterns within historical data. While these patterns can reveal profound truths about past performance and predict future probabilities with remarkable accuracy, they are inherently limited by the scope and fidelity of the data they consume. There will inevitably be instances where AI recommendations, though statistically sound based on the data presented, may not fully capture the nuances of a particular situation. This is where the indispensable role of human judgment comes into play, acting as a critical filter and an essential calibrator for AI-driven strategies.

Consider a scenario in which an AI system, having analyzed vast datasets on past marketing campaign performance, recommends a significant reallocation of budget to a specific digital advertising channel that has historically yielded high conversion rates. This recommendation is based on quantifiable metrics such as click-through rates, cost per acquisition, and return on ad spend. However, a seasoned marketing director might possess an intuitive understanding, gained from years of experience and direct customer interaction, that a subtle yet emerging shift in consumer sentiment is underway. This shift, perhaps not yet statistically significant enough to be detected by the AI, might indicate that the recommended channel, while historically strong, is approaching saturation or is beginning to alienate a growing segment of the target audience due to its perceived intrusiveness. In such a case, the human manager's judgment, informed by qualitative insights and a forward-looking perspective on brand perception, becomes crucial. They might decide to temper the AI's recommendation, perhaps by allocating a portion of the budget as suggested but reserving a substantial amount for experimental campaigns on newer, less proven channels that align with the evolving market sentiment. This approach leverages the AI's data-driven efficiency for established strategies while preserving the human capacity for strategic foresight and risk assessment in uncharted territory. The AI highlights the most probable path forward based on history, but the human manager navigates the broader landscape of possibilities and potential disruptions.

Bias, whether overt or subtle, is another critical consideration that necessitates human intervention. AI models are trained on data that reflects the world as it is, and unfortunately, the world often contains historical biases related to gender, race, socioeconomic status, or other demographic factors. If an AI system is tasked with recommending candidates for a new leadership role and historical hiring data shows disproportionate representation of a certain demographic in leadership positions, the AI might inadvertently perpetuate this bias. For instance, in current Fortune 500 corporate leadership roles, approximately 90% of CEOs are men and 10% are women. Racially, about 87% of these top positions are held by white executives, while Black, Hispanic, and Asian leaders represent roughly 1% to 3% each. If an AI uses this historical baseline, it could favor candidates who exhibit traits or backgrounds similar to those historically promoted, even if equally or more qualified candidates from underrepresented groups exist. In this context, the AI's recommendation, while seemingly objective due to its data-driven nature, is fundamentally flawed. A human decision-maker, armed with an understanding of ethical principles, diversity and inclusion goals, and a commitment to fairness, must critically scrutinize such recommendations. They need to look beyond the statistical correlations and actively seek to identify and mitigate any biases. This involves questioning the AI's output, probing the data it used, and actively seeking out diverse pools of talent that the algorithmic process might have overlooked. The human element provides the ethical compass, ensuring that decisions, especially those concerning human capital, are not only practical but also just and equitable, upholding the broader societal and organizational values.

Furthermore, AI systems often lack the deep contextual understanding that humans possess regarding organizational culture, internal politics, and the intangible factors that influence the success of any strategic initiative. An AI might recommend a particular course of action based on its projected ROI, but it cannot fully grasp the potential impact of that action on employee morale, interdepartmental relationships, or the organization's deeply ingrained norms and values. For example, an AI might suggest a restructuring that, while theoretically efficient, could lead to significant disruption, resistance from long-tenured employees, and a decline in team cohesion. The human leader, attuned to the organization's culture, understands that a gradual, more consultative approach might be necessary to achieve similar outcomes without alienating key stakeholders or undermining the company's cultural fabric. They can anticipate potential resistance, plan for change management, and communicate the rationale behind the decision in a way that resonates with the organization's values. This ability to navigate the human element of an organization—the emotions, relationships, and unwritten rules—is something AI currently cannot replicate. The AI provides the 'what' and the 'how' based on data, but the human leader brings the 'why' and the 'how best for us,' integrating the AI's insights into the complex tapestry of organizational reality.

The "Gravity of Command" implies a responsibility that extends beyond mere efficiency or profitability. It encompasses the ethical implications of decisions, the long-term sustainability of strategies, and the well-being of the people affected by those decisions. AI can be an unparalleled tool for enhancing efficiency and identifying opportunities, but it cannot bear the weight of command. That responsibility remains firmly with human leaders. This means critically evaluating AI outputs, not as immutable directives, but as valuable inputs into a more sophisticated decision-making process. It involves asking probing questions: Does this recommendation align with our long-term vision? Are there potential unintended consequences that the AI hasn't accounted for? Does this decision reflect our core values and ethical commitments? What is the human cost, or benefit, of this course of action?

Consider implementing a new AI-powered customer service system. The AI might recommend a standardized, script-based approach to handle customer inquiries, designed to maximize efficiency and minimize resolution times. From a purely data perspective, this might appear optimal, showing a reduction in average handling time and an increase in the number of queries processed per hour. However, a human manager, understanding the company's commitment to exceptional customer service and the importance of building strong customer relationships, might recognize that this approach could lead to impersonal interactions, customer frustration, and a decline in customer loyalty over time. They would understand that while efficiency is important, the quality of the customer experience is paramount. Therefore, they would likely intervene, perhaps by modifying the AI's parameters to allow for more flexibility, empowering human agents to deviate from scripts when necessary, and ensuring that complex or emotionally charged issues are escalated to human representatives. This is not about rejecting the AI's insight into efficiency, but about augmenting it with a deeper understanding of customer value and brand identity. The AI provides a data-driven benchmark for efficiency, but human judgment defines what constitutes truly effective customer engagement.

Furthermore, the dynamic nature of business environments means that AI models, trained on past data, can sometimes lag behind the pace of change. Market disruptions, unforeseen global events, or disruptive innovations can render historical patterns less predictive. In such scenarios, human intuition and experience become even more vital. For instance, during the early stages of the COVID-19 pandemic, many AI systems trained on pre-pandemic data would have struggled to accurately predict shifts in consumer behavior, supply chain vulnerabilities, or demand for certain products and services. Leaders who possessed a strong understanding of historical parallels, an ability to synthesize disparate pieces of information, and a willingness to act decisively in the face of uncertainty were able to adapt more quickly. They could supplement the AI's potentially outdated projections with their own informed assessments of the emerging landscape. This might involve making bold, albeit data-scarce, decisions to pivot business models, invest in new technologies, or rapidly reconfigure supply chains, guided by a combination of historical wisdom, current observations, and strategic foresight that transcends algorithmic prediction.

The art of balancing AI recommendations with human judgment is, therefore, a continuous process of calibration and integration. It requires fostering a culture where AI is seen as a mighty co-pilot, not an autonomous captain. Leaders must be adept at interpreting AI outputs, understanding their limitations, and knowing when to trust and when to question. This involves developing a critical mindset, demanding transparency in AI models, and advocating for the inclusion of qualitative data and ethical considerations alongside quantitative metrics. The ultimate goal is not to automate decision-making entirely, but to create a symbiotic relationship where AI amplifies human intelligence, enabling leaders to make decisions that are not only more informed and efficient but also more strategic, responsible, and aligned with the enduring values and aspirations of the enterprise. It is in this deliberate interplay between algorithmic insight and human discernment that actual strategic command and control are realized, ensuring that technology serves humanity's best interests and organizational objectives.

*****

The Strategic Radar Check

- **The "Weak Signal" Audit:** Ask your AI tool or data team to list three trends in your industry that are declining in volume and three that are accelerating in the last 90 days. The answers might surprise you.
- **The Devil's Advocate:** Take your current three-year strategy plan. Ask an AI: "Based on current tech trends, what are the top three risks that could cause this plan to fail?"
- **The ROI Calculator:** Review your top AI initiative. Have you defined a specific "Kill Switch"? For example: "If we don't see a 5% efficiency gain by Month 3, we pivot." If not, define it today.

6. Navigating Ethical Dilemmas in AI Leadership

The Ghost in the Machine

"It's not just efficient, Bob. It's a machine. It doesn't have an ego. It doesn't have bad days. It just picks the best talent."

Alice was glowing. The new AI recruitment pilot was processing resumes 500 times faster than the HR team. It had shortlisted ten candidates for the latest Senior Architect role in seconds. Bob picked up the shortlist. He read the names. He read the universities. He read the previous employers. He frowned.

"Alice," Bob said quietly. "These ten people... they all look exactly like me."

Alice blinked. She looked at the data. It was true. The AI had selected candidates who mirrored the demographics, educational background, and career paths of the company's current leadership—mostly male, primarily from three specific universities, and mainly from one particular socioeconomic bracket.

"The AI isn't biased," Alice defended, though her voice wavered. "It's trained on ten years of our own historical hiring data. It's looking for patterns of success."

"Exactly," Bob said, putting the list down. "It learned our history. And our history has flaws. If we use this tool, we aren't building the future; we are cloning the past. And we are automating discrimination at scale."

This is **Navigating the Grey**. In the leadership of AI, the most complicated problems are not technical; they are ethical. AI systems act like mirrors—reflecting the data they are fed. If that data contains the "ghosts" of historical bias, the machine will not only replicate that bias but amplify it with terrifying efficiency.

*****

The Trap of Algorithmic Bias

The challenge Bob identified is known as **Algorithmic Bias**. It is one of the most pervasive risks in AI leadership because it is often invisible. Unlike a human manager who might make a rude comment revealing their prejudice, an AI operates in silence, presenting its output as a mathematical fact.

- **How it Happens:** AI models learn from data. If your historical data shows a disproportionate representation of a certain demographic in leadership roles, the AI learns to favor candidates who mirror that demographic.
- The Consequence: The algorithm codifies prejudice, presenting exclusion as "objective decision-making."
- **The Leader's Duty:** You cannot blame the algorithm. In the **Gravity of Command**, if your tool discriminates, you are discriminating. You must move from "trusting the output" to "interrogating the input."
- **Action:** Leaders must scrutinize data sources. Does this dataset represent the diversity we want, or just the diversity we had?

*****

The Black Box: Transparency vs. Trust

Later that week, a different crisis emerged. The new AI-driven credit risk model denied a loan to a long-standing vendor. The vendor, called Bob, was furious, asking why. Bob asked the data team. Their answer? "We don't know."

The model was a "Deep Learning" neural network—a **Black Box**. It had made a connection between thousands of variables to decide the vendor was high-risk, but no human could trace the specific logic path.

"That isn't good enough," Bob told the team. "I cannot look a partner in the eye and say, 'The computer said no.' Leadership requires accountability. If we can't explain the decision, we can't defend it."

The Solution: Explainable AI (XAI)

To navigate the grey, leaders must demand **Explainable AI (XAI)**.

- XAI refers to techniques that allow humans to understand the "why" behind an AI decision.
- **The Standard:** It is not enough to know the AI is accurate (it gets the correct answer). You must know it is reliable (it gets the right answer for the right reason).
- **Alice's Shift:** Alice implemented a tool called **LIME** (Local Interpretable Model-agnostic Explanations), which highlighted which factors heavily influenced the rejection. It turned out the AI was flagging the vendor because of their zip code—a proxy for racial bias. Bob overruled the AI immediately.

*****

[Sidebar] Gravity Pulls: The Ethical Compass

- Fairness: Ensure the system does not create discriminatory outcomes for specific groups.
- **Accountability:** Establish a clear human "owner" for every AI decision. The algorithm is never responsible; the leader is.
- Transparency: If you can't explain it, don't deploy it for high-stakes decisions.

*****

The "Human in the Loop" Protocol

How do we prevent these disasters? We establish a **Human-in-the-Loop** protocol for all high-stakes decisions.

The “Gravity” Rule for Automation:

- **Low Risk** (e.g., Sorting Spam): Full Automation allowed.
- **High Risk** (e.g., Hiring, Lending, Medical Diagnosis): AI provides the recommendation; **Human makes the decision**.

In the hiring case, Alice and Bob redesigned the workflow. The AI didn’t “select” the shortlist. It “scored” candidates based on specific skills (Python, PMP) and flagged them for human review. The demographic data was blinded (hidden) from the AI to prevent it from learning bad habits.

The advent of Artificial Intelligence (AI) in leadership presents an unparalleled opportunity to enhance decision-making, optimize operations, and drive innovation. However, this transformative power is accompanied by a complex web of ethical considerations that demand proactive identification and rigorous analysis. As leaders, our responsibility extends beyond the immediate benefits of AI to encompass the potential long-term societal and organizational impacts.

To navigate this terrain effectively, we must cultivate a keen awareness of the ethical risks inherent in AI deployment, moving from a reactive stance to embrace a posture of vigilant foresight. This requires equipping ourselves with a robust toolkit for identifying these potential pitfalls early in the project lifecycle and integrating ethical scrutiny into the very fabric of AI development and implementation.

*****

The Risk of Algorithmic Bias

One of the most pervasive and insidious ethical risks associated with AI is **algorithmic bias**. AI systems learn from data, and if that data reflects historical societal inequities, the AI will inevitably learn and perpetuate those biases.

For instance, consider an AI system designed to screen job applications. If historical hiring data shows disproportionate representation of a particular demographic in senior roles, the AI might learn to favor candidates who mirror that demographic. This is not a hypothetical concern. To provide historical context, as of 2023, while **Black individuals made up roughly 14% of the U.S. population**, they held only about **3.2% of senior leadership roles** at Fortune 500 companies. Similarly, **women held only 10.4% of CEO positions** in the same group. If an algorithm is trained on this "success" data without intervention, it codifies these disparities as objective criteria for the future.

Leaders must move beyond overt discrimination checks and systematically question data provenance. This involves identifying whether a dataset represents the diversity we *want* for the future or merely the diversity we *had* in the past. Techniques such as **fairness-aware machine learning** should be explicitly integrated into model training to mitigate these gaps.

*****

Data Privacy and Security

Closely related to algorithmic bias is the risk of **data privacy violations**. AI systems often require vast amounts of data, frequently including sensitive personal information. A breach of this data can lead to reputational damage, legal penalties, and a profound loss of trust.

Identifying these risks requires a comprehensive understanding of governance frameworks such as GDPR and **CCPA**. Before deployment, a thorough data impact assessment should be conducted to evaluate the risks of privacy breaches. Leaders must champion **Privacy by Design**, a principle that advocates for embedding privacy considerations into the architecture of AI solutions from the very beginning.

*****

The "Black Box" and Explainable AI (XAI)

The **lack of transparency**, often referred to as the "black box" problem, poses a significant challenge in high-stakes environments such as medical diagnosis and financial lending. When a leader cannot explain *why* an AI made a specific decision, accountability vanishes.

To address this, leaders should prioritize **Explainable AI (XAI)**. This refers to techniques that allow humans to understand and trust the outputs of machine learning. Achieving this involves:

- **Choosing Interpretable Models:** Using decision trees or linear regression when the complexity of deep learning isn't strictly necessary.
- **Post-hoc Explanations:** Utilizing tools like **LIME** (Local Interpretable Model-agnostic Explanations) or **SHAP** (SHapley Additive exPlanations) to highlight which specific features—such as credit history or age—contributed most to a prediction.

Ultimately, the goal of XAI is to ensure the system is not just accurate but reliable for the right reasons. In regulated industries, understanding the causal links behind a prediction is not just an ethical preference; it is often a legal necessity.

The potential for AI automation to displace jobs poses a significant ethical concern for the modern workforce. As AI capabilities advance, many tasks previously performed by humans—particularly those defined by predictability and repetition—face the risk of automation. While this transition can be disruptive, leaders have an ethical obligation to manage it responsibly by conducting honest assessments of which job functions are most susceptible. However, the strategic focus should shift from mere displacement to transformation and augmentation. AI can liberate human workers from tedious, manual, or low-level cognitive tasks, allowing them to pivot toward more strategic, creative, and interpersonal roles. Proactive planning is essential here, necessitating investments in reskilling and upskilling programs that cover AI literacy, data analysis, and system oversight. By fostering a culture of continuous learning and transparent communication, organizations can empower employees to transition into new roles, such as robot technicians or maintenance specialists in manufacturing, rather than being replaced by automated systems.

Beyond the immediate concerns of employment, a more subtle risk lies in the over-reliance on AI and the subsequent erosion of human judgment. As these systems become more sophisticated, there is a dangerous temptation to unquestioningly accept their recommendations, which can diminish critical cognitive skills and lead to an abdication of leadership responsibility. In complex or ethically charged situations, human leaders must remain the final authority. Mitigating this risk requires a mindset that views AI as a powerful tool to augment human intelligence rather than a substitute for it. Leaders must establish clear boundaries for AI autonomy, ensuring that human oversight is mandatory for critical decisions. Regular post-mortem analyses of AI-assisted choices can help identify where human intuition, such as a sales director's nuanced understanding of seasonal consumer moods or competitive shifts, was necessary to correct or guide an algorithm's data-driven but historically limited projection.

Ultimately, identifying AI-related ethical risks requires fostering a deep-seated culture of moral awareness that permeates every level of the organization. This responsibility is not confined to technical teams; every individual involved in the AI lifecycle must understand the implications of their work. This involves ongoing education on best practices and the establishment of clear ethical guidelines that are communicated and strictly followed. By embedding a vigilant mindset into the core of their AI strategy, leaders can steer their organizations toward responsible innovation. This journey is an ongoing process of calibration, where robust governance frameworks and actionable policies serve as the bedrock for technological advancement, in the "Gravity of Command," ethical implications are not a final checklist but an integral part of the organizational DNA, ensuring that every deployment of AI remains aligned with the enduring human values of fairness, accountability, safety, and transparency.

Moreover, fostering a **culture of psychological safety** is essential for encouraging teams to raise concerns about potential ethical breaches without fear of reprisal. Employees need to feel empowered to question AI outputs, report observed biases, or flag privacy concerns. This can be achieved through anonymous reporting channels, regular team discussions dedicated to AI ethics, and explicit reassurances from leadership that raising ethical issues is a sign of diligence and commitment to responsible practice, not a performance deficit. When an engineer on an autonomous vehicle team feels comfortable flagging a subtle anomaly in sensor data that could have safety implications—even if it means delaying a deployment—it serves as a testament to a healthy ethical culture.

The development and implementation of ethical guidelines and governance frameworks are foundational to responsible AI leadership, providing the structure to translate abstract principles into day-to-day practices. By meticulously defining principles, establishing accountability, and ensuring transparency, leaders build a foundation that mitigates harm and unlocks the full potential of AI, fulfilling the **"Gravity of Command."** This proactive, systemic approach ensures that innovation is tempered with integrity and that technological advancement never overshadows human-centric values. A leader's mandate is to be the ethical custodian of these tools, ensuring they serve humanity rather than compromise it.

The imperative to ensure fairness and actively mitigate bias is a foundational requirement for trust and societal acceptance. Bias in AI rarely stems from malicious intent; instead, it is often a byproduct of historical societal inequities encoded in training data. For example, a facial recognition system trained predominantly on lighter skin tones may exhibit significantly lower accuracy for darker-skinned individuals. In a 2018 study, error rates for darker-skinned females were as high as **34.7%**, while those for lighter-skinned males were just **0.8%**. Similarly, recruitment tools trained on historical data can perpetuate cycles of exclusion. Leaders must foster an environment in which the potential for bias is acknowledged from the outset and rigorously scrutinize data sources and collection methodologies to identify proxy variables that could lead to discriminatory outcomes.

Detecting bias is a complex, ongoing process requiring technical expertise and critical oversight. During data exploration, teams should perform **Exploratory Data Analysis (EDA)** to look for demographic imbalances across protected attributes. Techniques like **statistical parity**—where selection rates are equal across groups—can serve as an initial indicator, though true fairness requires more nuanced considerations. Leaders must also champion qualitative assessments, engaging ethicists to review data for subtle biases in context or sentiment that quantitative analysis might miss. Specialized bias detection tools and libraries should be used to analyze datasets for disparities in representation and potential correlations with sensitive attributes.

Once bias is detected, the next step is implementing **debiasing strategies**. This iterative process often begins with **preprocessing methods that** modify the training data before it is fed into the model. Techniques include re-sampling (oversampling underrepresented groups or undersampling overrepresented groups), re-weighting data points to prioritize underrepresented instances, or generating synthetic data to augment sparse representations. However, these methods must be applied with caution, as they can inadvertently obscure important information or introduce new biases if not carefully managed.

To further mitigate bias, leaders can implement **in-processing techniques** that modify the learning algorithm during training. This involves adding regularization terms to the model's objective function that penalize unfair outcomes, or using **adversarial debiasing**. In this setup, a primary network attempts to make predictions while a secondary "adversary" network tries to predict protected attributes (like race or gender) from the primary model's output. The primary model is then forced to adjust its internal logic until the adversary can no longer identify those attributes, ensuring the final predictions are invariant to protected data.

Alternatively, **post-processing methods** are applied after a model is trained. These methods adjust final predictions to satisfy fairness criteria, such as setting different decision thresholds for various demographic groups to ensure parity in outcomes. While effective for immediate correction, leaders must weigh these interventions carefully. They can sometimes mask underlying data issues rather than solving them, potentially impacting model interpretability or predictive accuracy. Choosing the right strategy requires a clear understanding of the trade-offs between accuracy and the specific ethical or legal requirements of the application.

Beyond technical fixes, a vital component of ensuring fairness is the **diversity of the teams** building these systems. Homogeneous teams are more likely to share blind spots, overlooking biases that affect marginalized communities. For example, a 2019 study by the National Institute of Standards and Technology (NIST) found that facial recognition algorithms were often **10 to 100 times more likely** to misidentify Black and Asian faces compared to Caucasian faces. This frequently stems from a lack of representation in both the training data and the development rooms. Leaders must cultivate cross-functional teams that include data scientists, ethicists, and social scientists from diverse backgrounds to provide a comprehensive understanding of societal impacts.

Fostering a culture that seeks out diverse perspectives is paramount. This includes creating safe spaces for constructive debate where dissenting opinions regarding fairness are valued. Leaders should implement regular "bias review" sessions that go beyond technical metrics to delve into the societal implications of AI decisions. When an engineer feels empowered to flag a potential fairness issue, it acts as an early warning system that protects the organization from future ethical and legal liabilities.

The practical application of these principles is most critical in high-stakes domains. In **recruitment**, AI tools must be audited to ensure that their features do not serve as proxies for protected characteristics. For instance, a model that penalizes "gaps in employment" might inadvertently discriminate against women who have taken maternity leave. In **customer service**, AI must be tested with diverse user groups to ensure it understands various accents and dialects, preventing the exclusion of non-native speakers.

Finally, in **resource allocation**—such as loan approvals or insurance risk assessments—leaders must be vigilant about "proxy variables." A model that determines creditworthiness might inadvertently penalize certain ethnic groups if it uses zip codes as a factor, given the historical context of residential segregation. Leaders must insist on transparency, mandate regular external audits, and provide a robust appeals process for individuals who feel an algorithm has maltreated them. By doing so, they ensure that AI serves as a tool for equitable progress rather than a mechanism for systemic exclusion.

Moreover, the commitment to fairness extends to the continuous monitoring of AI systems in production. Bias is not a static problem; it can emerge or evolve over time as data distributions shift or as users interact with the system in unforeseen ways—a phenomenon often referred to as **model drift** or **concept drift**. Leaders must establish robust monitoring frameworks that track AI system performance across demographic groups in real time. This includes setting up automated alerts for significant deviations in performance or fairness metrics. When a potential bias is detected, there must be a clear protocol for investigation, remediation, and re-deployment. This might involve retraining the model with updated data, adjusting debiasing mechanisms, or even temporarily turning off the system if the bias poses a significant risk.

The role of leadership in this endeavor is not merely to delegate but to actively champion the cause of fairness. This means allocating sufficient resources for bias detection and mitigation, providing ongoing training for development teams on ethical AI principles, and holding individuals and teams accountable for upholding fairness standards. Leaders must also be willing to make difficult decisions, even if that means slowing deployment timelines or foregoing certain functionalities when the risk of perpetuating bias is high. The long-term consequences of biased AI—eroded trust, reputational damage, legal liabilities, and societal harm—far outweigh any short-term gains. Ultimately, ensuring fairness is a continuous journey. By prioritizing equity, leaders build AI that serves as a force for good, enhancing lives and contributing to a more just society.

*****

The Imperative of Transparency and Explainability

As AI systems influence everything from our next purchase to loan approvals, a vital ethical dimension demands our attention: **transparency and explainability**. As AI decision-making becomes more impactful, the demand for understanding how these decisions are reached grows exponentially. This is not merely an academic exercise; it is foundational for building trust, ensuring accountability, and enabling effective governance. Leaders must champion these principles, moving beyond the "black box" phenomenon where internal workings are so complex that even their creators struggle to articulate the reasoning behind a specific output.

Transparency refers to the degree to which an AI system's processes are open to scrutiny, while explainability focuses on articulating those processes in a human-understandable manner. This bridge is essential in high-stakes domains. Consider the healthcare sector: an AI designed to diagnose diseases must be transparent so clinicians can understand what evidence the AI prioritized. If an AI misinterprets an image or misses a subtle symptom, explainability allows for the identification of the failure point, preventing recurrence and safeguarding patient safety. Without it, clinicians are merely following a decree, which is antithetical to the core principles of medical practice.

Similarly, in financial services, transparency is often a legal imperative. When an individual is denied a loan, regulations such as the Fair Credit Reporting Act (FCRA) in the U.S. and the GDPR in Europe require that consumers be provided with specific reasons for the denial. If the system is a "black box," providing such explanations becomes impossible. Leaders must prioritize models and techniques that allow for actionable insights, whether by selecting inherently interpretable models—like decision trees—or by employing post-hoc techniques on complex neural networks. The goal is to empower individuals to understand their standing and seek recourse if an assessment is erroneous or unfair.

The implications for AI leadership are substantial. It means shifting from a sole focus on predictive accuracy to a dual imperative that balances performance with interpretability. This shift requires a deeper understanding of the diverse landscape of explainable AI (XAI) techniques, which can be broadly categorized into two groups: intrinsic explainability and post-hoc explainability. Intrinsic explainability refers to models that are inherently interpretable by design, while post-hoc explainability involves applying methods to understand the decisions of already-trained models, regardless of their complexity.

Intrinsic explainability is often achieved by using simpler, more transparent models. Linear regression, logistic regression, and decision trees are classic examples. A linear regression model, for instance, provides coefficients that directly indicate the weight and direction of each input feature's influence on the output. Decision trees, with their clear branching logic, are intuitive to follow; one can trace the path from the root node to a leaf node to understand how a particular decision was made. While these models might not consistently achieve the peak performance of complex neural networks, their transparency can be invaluable in many applications.

Leaders must consider the trade-offs: is the marginal gain in accuracy from a deep neural network worth the loss of interpretability, given that understanding the decision process is paramount? In many cases, the answer will be no. Investing in research and development to explore the "sweet spot" between performance and interpretability in these simpler models is a strategic imperative. Furthermore, developing hybrid models that combine the power of complex architectures with intrinsically interpretable components can offer a viable path forward.

Post-hoc explainability techniques offer a way to peer into the workings of complex, opaque models. These methods do not alter the model itself but aim to provide insights into its behavior. One prominent family of techniques is based on feature importance. Methods like Permutation Feature Importance assess the impact of shuffling the values of a single feature on the model's performance; if performance drops significantly, that feature is deemed necessary. SHapley Additive exPlanations (SHAP) is another powerful approach that draws on cooperative game theory to assign a unique importance value to each feature for a given prediction. SHAP values can explain how much each feature contributed, positively or negatively, to a specific outcome, providing local explanations for individual predictions.

Another set of post-hoc techniques focuses on surrogate models. These involve training a simpler, interpretable model, such as a decision tree, to mimic the predictions of the complex "black box" model in a local region of the input space. Understanding the surrogate model provides insights into the behavior of the complex model in that specific context. Local Interpretable Model-agnostic Explanations (LIME) is a prime example, generating local surrogate models to explain individual predictions. For instance, if an AI flags a document for containing sensitive information, LIME could highlight the specific words or phrases that contributed most to that classification.

For AI leaders, understanding and judiciously applying these XAI techniques is crucial. It means fostering a culture where data scientists and AI engineers are not only proficient in building models but also in evaluating their explainability. This involves incorporating XAI metrics into model evaluation frameworks alongside traditional metrics such as accuracy, precision, and recall. It also means investing in the right tools and platforms that support the implementation and visualization of XAI methods. The ability to generate and present clear, concise explanations to non-technical stakeholders is a critical skill that leaders must cultivate within their teams.

Effective communication of AI processes and decisions is a direct consequence of transparency and explainability. It is not enough to have explainable models; leaders must ensure that these explanations are communicated effectively to all relevant parties. This involves tailoring the explanation to the audience. A data scientist might require detailed technical explanations, including feature attributions and sensitivity analyses, whereas a regulatory body might need a high-level overview of the model's logic and its adherence to fairness and compliance standards. An end user whose AI decision has impacted them requires a simple, clear, and actionable explanation that helps them understand the outcome and, if necessary, how to appeal or rectify the situation.

This communication challenge extends to the documentation of AI systems. Comprehensive documentation should not only detail the model's architecture, training data, and performance metrics, but also explain the chosen XAI methods, their limitations, and the rationale behind their use. Leaders should advocate for adopting documentation standards that prioritize clarity and accessibility, ensuring that knowledge about AI systems is not siloed within small teams but is readily available to those who need it for oversight, auditing, or improvement. The goal is to demystify AI, making it an understandable tool rather than an inscrutable force.

Building trust is perhaps the most significant outcome of prioritizing transparency and explainability. In an era where AI is increasingly integrated into critical societal functions, public trust is a fragile commodity. When AI systems are perceived as opaque or arbitrary, it breeds suspicion and resistance. Conversely, when organizations can clearly articulate how their AI systems work and provide understandable reasons for their decisions, they foster confidence. This trust is essential for the widespread adoption and acceptance of AI technologies. Leaders who champion transparency and explainability are not just adhering to ethical best practices; they are making a strategic investment in the long-term viability and societal acceptance of the AI solutions they develop.

Transparency and explainability also play a crucial role in enabling effective oversight and troubleshooting. When an AI system behaves unexpectedly or produces erroneous results, understanding its decision-making process is paramount for identifying the root cause. Is the issue with the training data? Is there a flaw in the model's logic? Has an unforeseen interaction occurred between different components? Explainable AI provides the diagnostic tools necessary to answer these questions. Leaders must establish clear protocols for investigating AI-related incidents and leverage XAI techniques to pinpoint the source of the problem. This iterative process of monitoring, diagnosing, and correcting is fundamental to building robust and reliable AI systems. Without explainability, troubleshooting becomes akin to searching for a needle in a haystack, often leading to prolonged downtime, increased costs, and a diminished ability to recover from errors.

Furthermore, regulatory landscapes are rapidly evolving to demand greater transparency and explainability from AI systems as governments and international bodies grapple with the societal impacts of AI. Legislation and guidelines increasingly mandate disclosure and justification for AI-driven decisions, particularly in areas such as hiring, lending, and public services. Leaders who proactively embrace transparency and explainability will find themselves better positioned to meet these regulatory requirements, avoiding costly penalties and reputational damage. This foresight can also provide a competitive advantage, as organizations known for their transparent AI practices may gain preferential treatment from regulators and earn greater trust from customers and partners.

Implementing explainable AI is not without its challenges. There is often a perceived trade-off between model performance and interpretability. Highly complex models, such as deep neural networks, can achieve state-of-the-art accuracy but are inherently difficult to explain. Conversely, simpler, more interpretable models may not always achieve the desired level of predictive power.

Leaders must navigate this delicate balance, making informed decisions based on the specific context and the criticality of the application. This might involve accepting a slight reduction in accuracy in exchange for significant gains in transparency, or investing in research to develop novel XAI techniques that bridge this performance-interpretability gap.

Another challenge lies in the very definition of "understandable." What constitutes a clear and meaningful explanation can vary greatly depending on the audience. A reason that is comprehensible to an AI expert might be baffling to a layperson. Leaders must therefore ensure that their teams have the skills to translate complex AI concepts into accessible language and to present information in a way that is relevant and actionable for different stakeholders. This requires a blend of technical acumen and strong communication skills, which leaders should actively foster and recruit for.

The ethical implications of explainability are also profound. While XAI can help detect and mitigate bias, it can also be misused. For instance, if an AI system is designed to be discriminatory, an explainable model could reveal the discriminatory logic, allowing bad actors to refine their approach and make their discrimination harder to detect. Leaders must be vigilant against such misuse, ensuring that XAI is used as a tool for ethical governance rather than as a means to circumvent ethical principles. This underscores the importance of a holistic approach to AI ethics, integrating transparency and explainability with fairness, accountability, and robust governance frameworks.

In conclusion, as AI leadership evolves, the commitment to transparency and explainability must become a cornerstone of responsible AI development and deployment. It is the key to unlocking trust, ensuring accountability, and navigating the complex ethical dilemmas that arise from increasingly autonomous decision-making systems. By investing in XAI techniques, fostering clear communication strategies, and understanding the inherent trade-offs, leaders can move beyond the black box, building AI systems that are not only powerful and efficient but also understandable, trustworthy, and ultimately, beneficial to society. This proactive embrace of transparency and explainability is a hallmark of mature, ethical AI leadership, paving the way for a future in which AI serves humanity with clarity and integrity.

The increasing integration of artificial intelligence into critical business operations, while offering unprecedented opportunities for efficiency and innovation, simultaneously introduces new complexities and risks. As these systems become more autonomous and influential, the imperative to establish robust mechanisms for accountability and remediation becomes not just a best practice but an absolute necessity. The "Gravity of Command" in AI leadership means acknowledging that with great power comes great responsibility, and this extends to the potential failures of the very systems we deploy. When an AI system falters, produces erroneous outputs, or causes unintended harm, the organization and its leadership must be prepared to respond swiftly, diligently, and with a clear commitment to rectifying the situation. This involves moving beyond mere identification of issues to the proactive establishment of frameworks that ensure failures are understood, addressed, and prevented from recurring.

A fundamental component of accountable AI leadership is the development of comprehensive incident response plans tailored explicitly for AI-related failures. These plans should not be generic IT support protocols, but detailed blueprints designed to address the unique characteristics of AI systems. The initial step within such a plan is the meticulous investigation of root causes. This requires a multidisciplinary team, potentially including data scientists, AI engineers, ethicists, legal counsel, and domain experts, depending on the nature of the failure. The investigation must delve beyond surface-level symptoms to uncover the underlying factors.

Was the failure due to flawed data, an algorithmic bias that was not adequately addressed, a misinterpretation of context, a cyber-attack, or an unexpected interaction between different AI components? The granularity of the investigation is key; for instance, if an AI-powered customer service chatbot provides incorrect information, the investigation must determine if the error stemmed from a recent update to its knowledge base, a misunderstanding of a specific query's nuance, or a fundamental limitation in its natural language processing capabilities. For an AI used in predictive maintenance, a false positive flagging a component as faulty might be traced to sensor drift, insufficient training data for a specific operating condition, or changes in environmental factors affecting the machinery. The process of root cause analysis for AI failures often necessitates sophisticated debugging tools, model introspection techniques, and the ability to reconstruct the system's state and inputs leading up to the incident.

Following the identification of the root cause, a critical element of the remediation process is effective and transparent communication with all affected parties. This includes internal stakeholders, such as employees who rely on the AI for their work, as well as external parties, such as customers, partners, or regulatory bodies, whose error the AI's error may have directly impacted. The communication strategy must be tailored to the audience and the severity of the failure. For customers experiencing financial repercussions due to an incorrect AI-driven financial assessment, the communication should be empathetic, clear about the error, and outline the steps being taken to correct the situation and provide restitution.

For employees whose workflow was disrupted, communication should focus on the resolution timeline and any interim measures.

Regulatory bodies, especially those in heavily regulated industries like finance or healthcare, will require detailed reports, often within strict legal timelines, outlining the nature of the failure, the investigative findings, and the corrective actions implemented. This transparency builds trust and demonstrates a commitment to responsible AI governance, even in the face of adversity. The absence of clear, timely communication can exacerbate the negative impact of an AI failure, leading to reputational damage and eroding stakeholder confidence.

The implementation of corrective actions is the linchpin of the remediation process. These actions must directly address the identified root cause and aim to prevent recurrence. If the failure was due to biased training data, corrective actions might include augmenting the dataset with more representative samples, employing advanced bias-detection and mitigation algorithms, or retraining the model entirely. If the issue stemmed from a specific algorithm's vulnerability, the remediation might involve re-engineering the algorithm, implementing more robust validation checks, or switching to an alternative model architecture known for its resilience.

For instance, an AI system used for resume screening that systematically downranked candidates from certain demographic groups would require not only the identification of the bias in its training data or algorithmic weighting but also a fundamental overhaul of the screening process. This could involve retraining the model with de-biased data, implementing fairness-aware machine learning techniques, or even reverting to a more human-centric review process for sensitive stages, with AI acting solely as an initial filter for clearly unqualified candidates.

Corrective actions also include ongoing monitoring and re-evaluation of the AI system. Deployment is not the end of the AI lifecycle; it is merely a transition to a new phase of continuous improvement and oversight. Post-remediation, it is crucial to implement enhanced monitoring to detect subtle signs of recurring issues or new problems that may emerge. This involves setting up more sensitive anomaly-detection systems, conducting regular audits of AI outputs against defined benchmarks, and establishing feedback loops with users and affected parties to capture real-world performance deviations. The goal is to adopt a proactive posture in which potential issues are identified and addressed before they escalate into significant failures. This could involve A/B testing new versions of algorithms or data pipelines that have undergone remediation, meticulously tracking their performance and fairness metrics against established baselines.

Furthermore, the "Gravity of Command" mandates that leadership actively foster a culture that encourages reporting AI failures rather than penalizing them. Employees should feel empowered to raise concerns about AI behavior without fear of retribution. This requires establishing clear channels for reporting, such as dedicated AI ethics committees, internal helplines, or anonymous reporting systems. Training programs should equip employees with the knowledge to identify potential AI issues and understand the reporting procedures. Leaders must champion this open communication, demonstrating through their actions that learning from mistakes is a vital part of the AI development and deployment journey. This psychological safety is paramount; without it, potential failures might be hidden, allowing them to fester and cause greater harm later.

The legal and regulatory implications of AI failures are also a critical consideration for remediation. Depending on the jurisdiction and the nature of the AI system, an organization may face significant fines, legal liability, or mandated operational changes. Proactive development of remediation plans that align with evolving legal frameworks, such as GDPR, CCPA, or emerging AI-specific regulations, can mitigate these risks. This involves close collaboration with legal counsel to ensure that incident response protocols include provisions for data breach notification, regulatory reporting, and compliance with data protection laws. For example, if an AI system handling personal health information suffers a data breach due to a vulnerability, the remediation plan must immediately trigger protocols for notifying affected individuals and relevant data protection authorities, as mandated by regulations such as HIPAA or GDPR.

Beyond immediate technical and operational fixes, remediation also involves a strategic reassessment of the AI system's overall design, purpose, and ethical alignment. Was the AI system deployed in a context for which it was not truly suited? Were the initial risk assessments and impact analyses sufficiently thorough? Leaders must engage in post-failure reviews that go beyond technical details to examine the strategic decisions that led to the AI's deployment. This reflective practice is essential for learning and for ensuring that future AI initiatives are grounded in a more robust understanding of potential risks and ethical considerations. For instance, if an AI system intended for internal efficiency gains inadvertently leads to significant job displacement or increased workload for a specific department, the remediation might involve not just technical adjustments but a broader strategic review of the AI's role in the organizational structure and workforce planning.

Moreover, the concept of accountability in AI leadership extends to the establishment of clear governance structures. This means defining who is ultimately responsible for the AI system's performance, ethical implications, and failures. Is it the data science team, the product manager, the business unit leader, or a combination? While AI systems are complex, the human oversight and decision-making processes surrounding them must be clearly delineated.

This clarity of responsibility ensures there are designated individuals or teams to initiate and oversee remediation, preventing ambiguity and delays when incidents occur. This might involve establishing an AI governance board with defined oversight responsibilities for all AI deployments, mandating pre-deployment risk assessments, and assigning ownership for ongoing monitoring and incident response for each AI application.

In conclusion, establishing accountability and robust remediation processes for AI failures is not an optional add-on; it is an intrinsic component of responsible AI leadership. It requires proactive planning, thorough investigation, transparent communication, decisive corrective action, and a commitment to continuous learning and improvement. By integrating these principles into the organizational DNA, leaders can navigate the inevitable challenges that arise from AI deployment, ensuring that the "Gravity of Command" is exercised not only to harness the power of AI but also to mitigate its potential harms, thereby fostering trust and ensuring the ethical and sustainable advancement of artificial intelligence. This diligent approach reinforces the ethical commitment to both internal and external stakeholders, solidifying the organization's reputation as a responsible steward of advanced technologies.

Case Study: The "Rebellious" Algorithm

To understand the stakes, look no further than the launch of Grok, an AI developed with the intent of being "edgy" and less constrained than its competitors. While the goal was to avoid the "woke mind virus" (as its creators put it), the lack of traditional ethical guardrails led to immediate, catastrophic failure modes. Within weeks of its release, users found that with simple prompting, the AI would generate conspiracy theories, endorse antisemitic tropes, and in one disturbing instance, praise historical dictators as "effective" leaders for "decisive action."

Why did this happen?

1. **Data Pollution:** The model was trained on real-time social media data (X, formerly Twitter), which contains high volumes of unfiltered hate speech and bias.
2. **The "Cool" Factor vs. Safety:** In an attempt to give the AI a "personality," the developers lowered the safety thresholds that usually prevent toxic output. They prioritized engagement over ethics.

The Leadership Lesson: Grok did not "decide" to be offensive. It simply held up a mirror to the raw data it was fed, without filtering. As a leader, if you remove the guardrails to move faster or seem more "authentic," you are not innovating; you are exposing your brand to the raw, unrefined bias of the internet.

*****

Gravity Pulls: The Ethical Compass

- **Fairness:** Ensure the system does not create discriminatory outcomes for specific groups.
- **The "Grok Test":** Before deploying a model, ask: If this AI repeats the worst thing in its training data, will it destroy our reputation? If the answer is yes, you need better filters.
- **Accountability:** Establish a clear human "owner" for every AI decision. The algorithm is never responsible; the leader is.

*****

This example serves as a stark reminder: An AI without ethical brakes is not a "free thinker"—it is a liability.

Summary: The Moral Steward

Ethical AI leadership is not about stopping innovation. It is about steering it. Bob taught Alice that efficiency without ethics is just high-speed destruction. Alice taught Bob that with the right tools, such as XAI and bias detection, they could build a system that was actually fairer than human recruiters ever were.

In the next chapter, Communication and Stakeholder Management, we will look at how to explain these complex realities to the board, the team, and the customer without losing their trust.

*****

Mars Rover: Immediate Actions

The Ethical Risk Assessment Checklist: Before launching any AI pilot, ask these four questions:

- [] **Data Check**: Does the training data reflect the world we want to build, or the biases of the past?

- [] **The "Why" Test**: If a customer asks, "Why did the AI do this?", can we answer them in plain English?
- [] **The Impact Scan:** Who could be harmed by this system if it fails? (Think beyond the user to society at large).
- [] **The Kill Switch**: Do we have a process to shut down the model if we detect bias immediately?

7. Communication and Stakeholder Management in AI Projects

The Universal Translator

"Alice, stop," Bob said, gently closing her laptop. "If you show the Board a confusion matrix and an F1-score of 0.92, they won't see success. They will see a math problem. And executives hate math problems they cannot solve."

Alice sighed. "But the F1-score proves the model works! It proves we reduced the false positives by 40%."

"To you, it proves the model works," Bob corrected. "To them, it is just noise. You need to translate. Do not tell them we reduced false positives. Tell them we saved the Customer Support team 5,000 hours of wasted phone calls. Do not tell them the accuracy is 92%. Tell them that nine out of ten customers will get an instant answer without waiting on hold."

Alice looked at her slides. She deleted the matrix. She typed: "5,000 Hours Saved."

This is the challenge of Communication and Stakeholder Management in the AI era. As leaders, we often stand on the edge of a chasm. On one side is the intricate, probabilistic world of AI development. On the other side is the business world, which speaks the language of ROI, risk, and timeline. Your job is to be the bridge. You must translate the sophisticated into the strategic.

The Three Languages of Leadership

To bridge this chasm effectively, you must be fluent in three distinct dialects, switching between them as needed for your audience.

1. The Executive Dialect (Strategic Alignment) Executives care about three things: Money, Risk, and Reputation. They do not care about the architecture of your neural network.

- **The Mistake:** Explaining how the AI works (e.g., "We used a transformer model with 7 billion parameters").
- The Fix: Explaining what the AI achieves (e.g., "This architecture allows us to process claims 50% faster, improving cash flow by Q3").

2. The Technical Dialect (Precision and Feasibility) When speaking to your data scientists and engineers, ambiguity is the enemy. Here, you must talk in specifics: data pipelines, latency requirements, and validation sets.

- **The Goal:** To ensure everyone understands the constraints. If you promise the Board "real-time processing" but tell the engineers "fast processing," you have created a failure gap.

3. The End-User Dialect (Usability and Benefit) For the people actually using the tool, the technology is irrelevant. They care about their workflow.

- **The Narrative:** Do not sell them "AI Automation." Sell them "Relief." Explain how the tool removes the boring parts of their job, such as data entry, so they can focus on the interesting parts, like solving problems.

*****

Gravity Pulls: The "So What?" Test

Before every status update or presentation, apply the "So What?" test to your key points.

- **Draft:** "We completed the data ingestion phase."
- **So What?** "We now have a clean dataset."
- **So What?** "We can begin training the model."

- **So What? (The Final Pitch)** “We remain on track to launch the pilot next month, which will begin generating revenue by Q4.”

*****

Managing Expectations: The “Magic Wand” Problem

The most considerable risk in AI communication is not technical failure; it is misaligned expectations. Because of media hype, stakeholders often view AI as a “magic wand”—a tool that works instantly and perfectly. When the V1 model hallucinates or fails to handle edge cases, stakeholders panic.

The “Evolutionary” Narrative: You must frame AI projects not as a “Launch” (static) but as an “Evolution” (dynamic).

- The Script: “The version we launch on Day 1 is the worst this system will ever be. It is designed to learn. We expect a 70% success rate in week one, growing to 90% by month three as it learns from your feedback.”

This reframing turns early bugs from “failures” into “lessons.” It prepares stakeholders for the iterative nature of AI, in which the system must encounter the real world to improve.

The Surround Sound Effect

Communication cannot be a one-time broadcast. It must be "Surround Sound"—a consistent message echoed across all channels until it becomes the culture. If Bob tells the board, "This is a pilot," but Alice tells the team, "We are going to production," the project will tear itself apart.

Creating the Echo Chamber:

1. **Consistent Vocabulary:** Define your terms. Does "MVP" mean "prototype" or "market-ready"? Define it once and enforce it.
2. **Regular Cadence:** Silence breeds conspiracy. In the absence of updates, people assume the project is failing. Even if the update is "We are still training the model," say it.
3. **The Feedback Loop:** Communication is two-way. Create safe channels for end-users to say, "This is not working," without fear. Their feedback is the fuel your AI needs to get better.

Bridging the chasm between the intricate world of artificial intelligence development and the diverse understanding of an organization's stakeholders is a cornerstone of successful AI project management. The very nature of AI, with its reliance on complex algorithms, vast datasets, and abstract computational processes, can often create a barrier to comprehension for those who are not steeped in the technical details.

As leaders, our role is to act as conduits, translating these sophisticated concepts into language that resonates with every level of the organization, from the executive suite to the front-line user. This translation is not about dumbing down the technology, but about illuminating its purpose, progress, and ultimate value in terms accessible and relevant to each audience.

The initial and perhaps most crucial step in effectively communicating complex AI concepts is to understand who you are speaking to. A detailed technical presentation for fellow data scientists will be vastly different from a high-level overview for a board of directors or an explanation of a new AI-powered feature for the sales team. For the executive leadership, the focus should invariably be on strategic alignment, business outcomes, and return on investment. They are concerned with how the AI initiative contributes to the company's overarching goals, its competitive advantage, and its financial health.

Therefore, instead of delving into the intricacies of a convolutional neural network's architecture, one might highlight how that architecture enables a more accurate predictive model that will increase sales forecasts by 15%, directly impacting revenue targets. The emphasis here is on the "what" and the "why" from a business perspective, rather than the "how" from a technical one. This requires a deep understanding of the business objectives and the ability to map the AI's capabilities directly to them.

When addressing the technical teams involved in the AI project, the communication can, and often should, be more granular. Here, clarity around the specific algorithms, data pipelines, model performance metrics, and potential challenges is paramount. However, even within technical teams, there can be diverse specializations. A machine learning engineer might understand the nuances of a specific algorithm, while a data engineer might be more focused on the infrastructure and data flow. Effective communication in this context involves fostering a shared understanding of the project's goals, ensuring that all team members understand how their specific contributions fit into the larger picture, and facilitating open discussion of technical hurdles and potential solutions. This often involves consistently using precise technical terminology and ensuring that all relevant parties understand it. Regular technical sync-ups, code reviews, and internal documentation become critical tools.

For end users or departmental teams who will directly interact with or benefit from the AI system, the communication should focus on usability, practical benefits, and how the AI will simplify or enhance their daily tasks. If an AI is designed to automate a reporting process, the explanation should focus on how it will reduce the time spent on manual data entry, improve accuracy, and provide insights more quickly, rather than on the natural language processing techniques used to extract information. Demonstrating the AI's value through tangible improvements in their workflow is key. This might involve user training sessions, interactive demonstrations, and the creation of user-friendly documentation that explains the AI's functionalities in simple, action-oriented terms. Case studies of how similar AI tools have benefited other users or teams can also be highly persuasive.

A fundamental strategy for simplifying complex AI concepts is to avoid jargon wherever possible. Terms like "hyperparameter tuning," "gradient descent," or "backpropagation" are likely to alienate and confuse non-technical audiences. Instead, use analogies and metaphors grounded in everyday experiences. For example, explaining an AI's learning process is like a child learning to identify different animals by being shown many pictures and told what each one is, or a chef learning to perfect a recipe through trial and error and feedback. When a technical term is unavoidable, define it clearly and concisely immediately after its first use, using simple language. The goal is to make the concept understandable without sacrificing accuracy. Think of it as translating from one language to another, ensuring the meaning is preserved while making it comprehensible to the target audience.

Visual aids can be incredibly powerful tools for demystifying AI. Charts, graphs, flow diagrams, and even simple infographics can help illustrate complex processes or data relationships more effectively than words alone.

For instance, a flowchart can visually represent the steps an AI takes to process a customer query, from initial input to the final response, making the workflow much easier to grasp. A graph showing the improvement in accuracy over time as a model is trained can provide a straightforward, data-driven narrative of progress. When presenting the impact of an AI system, visualizations that show before-and-after scenarios, or highlight key performance improvements, can be far more impactful than a table of numbers. These visuals should be clean, uncluttered, and directly relevant to the message being conveyed.

Storytelling is another potent technique for communicating AI's value. Humans are wired to connect with narratives. Instead of just presenting data points, weave them into a story that illustrates the problem the AI is solving, its development journey, and the positive impact it has achieved. This could be the story of how an AI-powered diagnostic tool is helping doctors identify diseases earlier, leading to better patient outcomes, or how an AI-driven recommendation engine is helping customers discover products they love, enhancing their shopping experience. These narratives humanize the technology, making its benefits relatable and memorable. The story should highlight the challenges overcome, the insights gained, and the ultimate positive transformation brought about by the AI.

Managing expectations is intrinsically linked to effective communication. AI is not magic; it is a sophisticated technology with limitations. It is crucial to be transparent about what the AI can and cannot do, especially in the early stages of development and deployment. Overpromising and under-delivering can quickly erode trust and lead to disappointment. When discussing the capabilities of an AI system, it is important to temper enthusiasm with realism.

For instance, if an AI chatbot is being developed, it is crucial to convey that it might not be able to handle every complex or nuanced query and that human intervention will still be necessary in certain situations. Clearly defining the scope and limitations of the AI and communicating them to all stakeholders helps set realistic expectations and prevent misunderstandings down the line.

Regular and consistent communication is vital. AI projects are often iterative, involving continuous learning, refinement, and adaptation. Providing stakeholders with regular updates, even if there is no groundbreaking news, helps to maintain engagement and build confidence. These updates can take various forms: weekly or bi-weekly email summaries, monthly review meetings, or dashboard updates that track key performance indicators. The frequency and format of these updates should be tailored to the audience's needs and preferences. The key is to keep stakeholders informed about progress, challenges, and any shifts in strategy. This proactive communication approach fosters a sense of partnership and shared ownership of the AI initiative.

When communicating progress, it's essential to translate technical milestones into business achievements. For example, successfully completing a phase of data cleaning and preprocessing might not mean much to an executive, but framing it as "laying the foundation for a more accurate customer segmentation model that will enable personalized marketing campaigns" directly connects the technical work to a tangible business outcome. Similarly, achieving a specific accuracy threshold in a model could be translated into "a significant reduction in the rate of false positives, saving the company X dollars in wasted resources or missed opportunities." This "business translation" is a critical skill for AI project managers. It requires constantly asking "So what?" after every technical update.

Furthermore, creating feedback loops is essential for both improving the AI system and refining communication strategies. Actively soliciting feedback from stakeholders about the clarity of information, the perceived value of the AI, and any concerns they might have provides invaluable insights. This feedback can inform future communication efforts, helping to identify areas where explanations need to be clearer or where different approaches might be more effective. It also demonstrates that their input is valued and that the AI project is a collaborative endeavor. This can be done through surveys, informal conversations, or dedicated feedback sessions.

For user-facing AI applications, user acceptance testing (UAT) and pilot programs are invaluable opportunities for communication and feedback. By involving end users in testing the AI's functionality in a real-world setting, you can not only identify bugs and usability issues but also gauge their understanding and perceptions of the AI. The insights gained from UAT can then be used to refine the AI system and its accompanying communication materials before a full-scale rollout. Explaining the purpose of UAT and clearly defining the feedback needed from users can help users feel more invested in the process and more comfortable providing honest opinions.

When AI initiatives encounter challenges or setbacks, transparent communication is even more critical. Instead of hiding problems, leaders should address them directly, explaining the nature of the challenge, its potential impact, and the plan to overcome it. This honesty builds trust and demonstrates resilience. For instance, if a model's performance degrades unexpectedly, it is better to communicate the issue, the investigation underway, and the expected timeline for a fix, rather than letting stakeholders discover the problem through its consequences. The explanation should focus on the technical reasons for the setback in an understandable way, followed by the strategic steps being taken to rectify it. This might involve re-training the model with new data, refining the algorithms, or adjusting system parameters.

It is also important to consider the ethical implications of AI when communicating with stakeholders, particularly regarding data privacy, bias, and transparency. Depending on the audience, these discussions might need to be framed in terms of regulatory compliance, brand reputation, or customer trust. For example, when explaining how an AI system uses customer data, it's crucial to reassure stakeholders about the measures in place to protect that data and ensure its ethical use. Highlighting the steps taken to mitigate algorithmic bias and ensure fairness can be a key part of building confidence, especially for customer-facing applications or those impacting sensitive decisions. Communicating these ethical considerations proactively can prevent concerns from arising later and demonstrate a responsible approach to AI development.

Finally, celebrating successes is an integral part of effective communication. When an AI initiative achieves its goals or demonstrates significant value, it is vital to share that success broadly across the organization. This not only reinforces the positive impact of AI but also builds enthusiasm and encourages further adoption and investment. These celebrations should not just be about announcing a successful outcome but also about acknowledging the effort and collaboration that led to that success, thereby reinforcing the value of teamwork and the importance of AI in the organization's strategic vision.

This storytelling of success, backed by clear metrics, reinforces the "Gravity of Command" by demonstrating the tangible benefits of AI, making it naturally easier to secure future resources and strategic focus. The ability to articulate these successes in a clear, compelling, and audience-appropriate manner is a hallmark of effective leadership in the age of AI.

Formal presentations also remain a vital component of stakeholder communication, particularly for significant milestones, phase gates, or when seeking further investment or approvals. These presentations should be structured to tell a compelling story, guiding the audience through the project's journey. A typical structure might include:

1. **Introduction and Project Objectives:** A brief recap of the "why" behind the project and the strategic goals it aims to achieve.
2. **Progress Update:** A summary of work completed since the last update, highlighting key achievements and technical breakthroughs. This section should be balanced with a transparent discussion of any challenges encountered and how they are being managed.
3. **Demonstration of Value:** This is often the most crucial part. It involves showcasing tangible evidence of the AI's impact through data, case studies, or even live demonstrations if feasible. Quantifying benefits with clear financial or operational metrics is essential here. For example, if an AI

system is designed to optimize logistics routes, a presentation might include a simulation showing a reduced delivery time or fuel consumption compared to the previous method, backed by projected cost savings.

4. **User Feedback and Adoption:** Insights from end-users or pilot groups, demonstrating how the AI solution is being received and integrated into workflows. This human element adds credibility and reinforces the technology's practical utility.
5. **Future Roadmap and Next Steps:** An outline of planned activities, upcoming milestones, and any emerging opportunities or potential adjustments to the strategy based on learnings.
6. **Call to Action/Decision Required:** If the presentation is intended to solicit a decision or request further resources, this should be clearly stated with a supporting rationale.

The language used in reporting is as critical as the data itself. Technical jargon should be minimized or, at a minimum, clearly explained when communicating with non-technical audiences. The focus should always be on the "so what"—the business implications and strategic value derived from the AI project. For instance, instead of stating, "We have achieved a 92% F1-score on our image recognition model," a more impactful statement for business stakeholders might be, "Our AI system can now identify critical product defects with over 90% accuracy, significantly reducing manual inspection time and improving product quality control, which is projected to save $X annually in material waste and rework."

It's also important to acknowledge that the clarity and credibility of progress reporting directly influence the "Gravity of Command" in AI projects. When stakeholders have confidence in the project team's ability to track progress, identify and solve problems, and clearly articulate value, their willingness to commit resources, trust the strategy, and champion the initiative is significantly enhanced. Conversely, vague reporting, a lack of transparency about challenges, or a failure to demonstrate tangible business outcomes can erode confidence and lead to a withdrawal of support, even if the underlying technology is sound. Effective reporting, therefore, is an active process of building and maintaining this vital stakeholder trust.

When reporting on AI value, consider different levels of impact. There are direct, quantifiable benefits, such as cost savings from automation. Then there are indirect benefits, such as improved decision-making capabilities that might not have an immediate, easily measurable dollar value but contribute to long-term strategic advantage. Furthermore, consider the impact on organizational capabilities: has the project fostered greater data literacy, enhanced innovation culture, or improved cross-functional collaboration? Documenting and communicating these broader impacts, even if qualitative, adds depth to the value proposition. For example, a project that automated routine data entry might not only show direct labor cost savings but also highlight how it freed up employees to focus on higher-value analytical tasks, thereby enhancing the team's overall capacity for strategic insight.

To ensure reporting is consistently effective, consider establishing a "reporting framework" that outlines the types of information to be collected, the metrics to be tracked, the reporting channels, and the responsibilities for each.

This framework should be a living document, reviewed and updated periodically as the project evolves and stakeholder needs change. Regular feedback loops on the reporting process itself are also beneficial. Asking stakeholders, "Is this report clear?" or "What additional information would be helpful?" or "Is the current cadence of updates meeting your needs?" can help refine the communication strategy.

In conclusion, the subsection on Reporting Progress and Demonstrating Value emphasizes that ongoing, transparent, and value-centric communication is not an optional add-on for AI projects but a core component of their success. By diligently tracking progress, candidly addressing challenges, and consistently articulating the tangible business benefits in clear, data-driven terms, project managers can ensure stakeholders remain informed, engaged, and supportive. This continuous narrative-building, supported by well-crafted reports and dashboards, is essential for validating the project's strategic importance and securing the sustained commitment needed to harness the full potential of artificial intelligence. It is through this diligent reporting that the "Gravity of Command" is not only maintained but strengthened, guiding the AI initiative towards its intended impact and organizational goals.

The introduction of Artificial Intelligence into an organization is rarely a smooth, universally welcomed transition. Even with the most compelling business cases and the most robust technical solutions, human elements of change management often introduce friction. Resistance, whether overt or subtle, is a natural reaction to the unknown, to perceived threats, or to the disruption of established routines. Navigating this resistance and cultivating genuine buy-in requires a deliberate, empathetic, and strategic approach that goes beyond simply presenting data and highlighting potential benefits. It demands an understanding of the underlying causes of apprehension and a commitment to building trust and a shared vision.

The role of leadership in fostering buy-in cannot be overstated. Senior management must not only champion the AI initiative verbally but also visibly support its adoption. This includes allocating necessary resources for implementation, training, and ongoing maintenance. Leaders demonstrating their own willingness to engage with and learn about the AI systems, even at a high level, sends a powerful message. They can share their own experiences and how AI is impacting their strategic decision-making.

This visible endorsement from the top helps legitimize the initiative and signals its strategic importance to the entire organization. When leaders prioritize AI adoption and integrate it into their own workflows, it validates the investment and encourages broader adoption.

Acknowledging and celebrating early successes, no matter how small, is vital for building momentum and reinforcing positive perceptions. When a pilot program yields positive results or when an individual or team successfully adopts and benefits from a new AI tool, these achievements should be recognized and communicated widely. This reinforces the value proposition and encourages others to follow suit. Recognition can take various forms, from internal newsletters and team meetings to formal awards or acknowledgments. The goal is to create a positive feedback loop where demonstrated success breeds further adoption and enthusiasm.

It is also important to maintain a realistic perspective on AI's capabilities and limitations. Over-promising and under-delivering is a surefire way to erode trust and foster long-term resistance. Be upfront about what the AI can and cannot do, and set clear expectations for the timeline to achieve specific outcomes. If there are known limitations or potential challenges, address them proactively rather than waiting for them to become apparent. For instance, if a sentiment analysis AI is being deployed for customer feedback, be clear that it might not perfectly capture sarcasm or highly nuanced language, and that human review may still be necessary for critical feedback. This honesty builds credibility and manages expectations, preventing disappointment and subsequent backlash.

Finally, establishing clear channels for feedback and continuous improvement is essential. Once an AI system is deployed, the process of gathering user feedback and making iterative improvements should be ongoing. This demonstrates a commitment to user experience and ensures the AI solution continues to evolve to meet the organization's and its employees' needs. Regularly soliciting input on what is working well, what could be improved, and what new capabilities would be beneficial can help maintain engagement and ownership.

This feedback loop also provides valuable insights for future AI initiatives. By actively involving users in the evolution of technology, organizations can foster a culture in which AI is seen as a valuable partner in achieving collective goals, rather than a source of disruption or apprehension. This continuous dialogue and commitment to refinement are key to transforming initial skepticism into sustained support and, ultimately, driving the successful integration of AI across the enterprise.

*****

Mars Rover: Immediate Actions

The Difficult Conversation Scripts: Use these templates when the news is not good.

- **Scenario: The Model is Not Ready**
  - **Do Not Say:** "It is broken."
  - **Do Say:** "We have encountered a complexity in the data integration that requires an extra sprint to resolve. We are prioritizing accuracy over speed to ensure the pilot data is reliable."
- **Scenario: Fear of Replacement**
  - **Do Not Say:** "Do not worry about it."
  - **Do Say:** "This tool is designed to automate the [X] task, which frees you up to focus on [Y], where human judgment is critical. We are investing in training so you can master this new tool."

*****

Summary

Alice's presentation to the Board was a success. She did not show the confusion matrix; she showed a graph of Time Saved. She did not promise a magic wand; she promised a Learning Engine. By translating the tech into value, she did not just get approval—she achieved alignment.

In the next chapter, **Future-Proofing Your Leadership**, we will look at how to sustain this success by building resilience and a legacy that outlasts the technology itself.

8. Future-Proofing Your Leadership in the AI Era

The Gravity Leader

"You know, Alice," Bob said, looking at his screen. He was not looking at email; he was looking at a Python script he had just debugged with the help of an AI agent. "I used to think my value was knowing all the answers. Now I realize my value is knowing which questions haven't been asked yet."

Alice smiled. She was holding a physical notebook, sketching a conflict-resolution map for two feuding stakeholders. "And I used to think data was the only truth. But data couldn't fix the trust issue between Marketing and Engineering. Only empathy could do that."

They had traded places, in a sense. Bob, the master of human connection, was embracing continuous learning to stay relevant. Alice, the master of the machine, was embracing emotional intelligence to become holistic. They had become Gravity Leaders.

This is the essence of Future-Proofing. In an AI-native world, the leaders who survive are not those who cling to the past through pure intuition, nor those who mindlessly rush into the future through pure automation. They are the ones who generate the gravity required to hold both worlds together.

Resilience: The Art of Bouncing Forward

The pace of AI innovation is not slowing down. What is cutting-edge today—like Generative AI—will be legacy tech in three years. If your leadership style is rigid, you will snap; if it is fluid, you will adapt.

The "Odd" Philosophy

In a world where AI can standardize average performance for everyone, "average" is no longer a safety net; it is a commodity. To lead effectively, you must diverge from the mean. As the saying goes, "You have to be odd to be number one."

What does it mean to be "odd" in the AI era?

1. **Be Oddly Human:** When everyone else is automating their emails, be the leader who picks up the phone. High-touch communication becomes a premium service in a low-touch world.
2. **Be Oddly Curious:** While others are optimizing their current roles, spend 20% of your time exploring tools that might replace your role. Do not wait for disruption; invite it in for coffee.
3. **Be Oddly Slow:** In a culture of instant algorithmic answers, be the one who pauses to ask, "Is this the right problem to solve?" Slow thinking is the ultimate strategic advantage against fast AI.

The Leadership Legacy: Beyond the Deliverable

Bob and Alice realized that their project was not just about launching software. It was about building a team that could launch anything, no matter how the technology changed.

Building "Anti-Fragile" Teams

Nassim Nicholas Taleb introduced the term antifragile to describe systems that don't merely survive stress — they improve because of it.

- **Fragile Team:** Breaks when the AI tool fails or the requirements change.
- **Resilient Team:** Survives the change and returns to the baseline.
- **Anti-Fragile Team:** Uses the disruption to learn a new way of working and comes out ahead.

Antifragility is not about loving chaos. It's about designing teams that treat uncertainty as feedback. Small shocks become practice runs.

Mistakes become data. Change becomes acceleration. In the AI era—where tools evolve monthly, and assumptions expire quickly—resilience is no longer enough. The teams that win are the ones that grow through volatility, not despite it.

How to Build It:

- **Democratize the Tools:** Do not centralize AI access in a small elite group. Give the intern the same experimental sandbox as the Chief Architect. When more people can test ideas, you increase the number of small experiments — and small experiments produce small failures. Small failures prevent catastrophic ones. Innovation should bubble upward, not wait for permission.
- **Celebrate "Intelligent Failure":** Not all failure is equal. Reckless failure is waste. Intelligent failure is data. If an AI experiment fails but reveals a flaw in your assumptions, you just paid tuition for future speed. A team afraid of failure becomes conservative. A conservative team becomes fragile. An antifragile team learns in public and improves with every iteration.

*****

Light Year Lesson: The Legacy Check

"Your legacy isn't the project you delivered. It's the people you developed. Ten years from now, nobody will remember if the migration happened on Tuesday or Thursday. They will remember that you taught them how to navigate uncertainty without losing their soul." — Bob's Retirement Speech (Draft)

*****

Mars Rover: Immediate Actions

The AI Leadership Readiness Assessment: Use this diagnostic to see if you are future-proofed.

Pillar	The Question	The Action
Strategic Understanding	Can you explain to a 5-year-old how AI helps your specific business make money?	If not, go back to the "Moonlight" chapter and map your AI use cases to ROI [8].
Team Development	Have you hired or trained someone this year who has a skill *you* don't understand?	If you are the most intelligent person in the room, you are in the wrong room. Hire for *curiosity*, not just keywords [9].
Ethical Governance	Do you have a "Kill Switch" for your AI projects if they show bias?	If not, establish an ethical review board today. Don't wait for a PR crisis [10].
Adaptability	When was the last time you changed your mind because of data?	Practice being "odd." Challenge your own assumptions publicly [11].

The landscape of artificial intelligence is not a static entity; it is a perpetually shifting terrain, characterized by rapid innovation and the constant emergence of novel capabilities. For leaders aiming to future-proof their organizations and their own leadership effectiveness, anticipating the next wave of AI advancements is paramount. This requires cultivating a proactive and vigilant approach to understanding the trajectory of AI research, development, and its potential applications across industries. It is about moving beyond reactive adoption and embracing a strategic posture of foresight, enabling organizations not only to adapt to change but also to actively shape their future in alignment with technological evolution.

Staying informed about emerging research and industry developments is the bedrock of this anticipatory leadership. This involves establishing deliberate channels for continuous learning and intelligence gathering. Beyond passive information consumption, leaders must foster a culture of active exploration within their organizations. This might involve setting up internal "AI watch" groups or task forces whose mandate is to monitor technological advancements, assess their relevance to the business, and report back on potential opportunities and threats. These groups can act as an internal intelligence unit, sifting through vast amounts of information to identify the signals that matter most to the organization's strategic objectives.

Encouraging employees at all levels to share interesting AI developments they encounter, whether through internal forums, newsletters, or informal discussions, can create a distributed network of awareness. Recognizing and rewarding employees who bring forward valuable insights from the AI landscape can incentivize this proactive engagement.

Furthermore, understanding the underlying drivers of AI innovation is crucial for anticipating future advancements. Several key areas are consistently pushing the boundaries of what AI can achieve. One such area is the continued evolution of deep learning architectures and training methodologies. Innovations in areas like self-supervised learning, few-shot learning, and more efficient transformer models are enabling AI systems to learn with less labeled data and to perform more complex tasks.

For leaders, this means recognizing that the limitations of current AI systems, often related to data requirements or generalization capabilities, are likely to diminish over time. This opens up new possibilities for applying AI in domains where data is scarce or the tasks are highly nuanced.

Another significant frontier is the development of more robust and explainable AI (XAI). As AI systems become more integrated into critical decision-making processes, the demand for transparency and understanding of how these systems arrive at their conclusions will only grow. Research into techniques that allow for clearer interpretation of model behavior, identification of biases, and assessment of confidence levels is vital. Leaders should be aware that the ability to explain AI decisions will become a competitive differentiator, particularly in regulated industries. Investing in and advocating for XAI development will not only build trust but also mitigate risks associated with opaque decision-making.

The convergence of AI with other emerging technologies also presents a fertile ground for future advancements. The synergy between AI and the Internet of Things (IoT) is creating a new generation of intelligent, connected devices capable of real-time data analysis and autonomous action. This fusion is driving innovation across areas such as smart manufacturing, predictive maintenance, and personalized healthcare.

Leaders need to consider how this convergence can create new business models, enhance operational efficiencies, and provide novel customer experiences. Similarly, the interplay between AI and edge computing enables sophisticated AI processing closer to the data source, reducing latency and enhancing privacy. This development is particularly relevant for applications requiring immediate responses, such as autonomous vehicles or real-time anomaly detection in industrial settings.

The realm of generative AI, which has seen rapid progress in recent years with models capable of creating text, images, music, and even code, represents another transformative wave. While current applications are often focused on creative content generation, the underlying principles have far-reaching implications. Generative AI could revolutionize product design, accelerate scientific discovery through hypothesis generation, and enable more sophisticated simulations. Leaders should actively explore how these capabilities can be integrated into their R&D processes, marketing efforts, and even customer service interactions to foster innovation and efficiency.

The ethical implications and potential for misuse of generative AI also warrant careful consideration and proactive policy development.

Anticipating the ethical and societal implications of AI is as crucial as understanding its technical advancements. As AI systems become more pervasive, questions around bias, fairness, privacy, and accountability will become increasingly prominent. Regulatory frameworks are likely to evolve, and public perception will play a significant role in the adoption and acceptance of AI technologies.

Leaders must stay abreast of these evolving discussions and proactively consider the ethical dimensions of their AI strategies. This includes establishing robust ethical guidelines, ensuring diverse representation in AI development teams, and actively engaging with stakeholders to address societal concerns. A forward-thinking leader will not only seek to leverage AI for business advantage but will also strive to do so responsibly and equitably.

Developing the capacity for strategic foresight also involves understanding the potential for disruptive AI applications. Sometimes, it is not the incremental improvements in existing AI capabilities that cause the most significant shifts, but rather entirely new ways of applying AI that fundamentally alter industries. This could involve AI systems that automate complex decision-making currently performed by highly skilled professionals, AI-powered platforms that disintermediate existing value chains, or AI-driven insights that unlock entirely new markets. Identifying these potential disruptors requires a willingness to challenge existing assumptions about how business is done and to look beyond established industry boundaries.

Scenario planning, a technique in which leaders explore multiple plausible future states and their potential impacts, can be a valuable tool in this regard. By imagining various future scenarios, including those driven by radical AI advancements, organizations can better prepare for a range of potential outcomes and develop more resilient strategies.

Moreover, fostering a mindset of continuous learning and adaptation within the organization is key to navigating an era of perpetual AI evolution. This means not only investing in individual employees' skills to understand and work with AI, but also building organizational agility. An agile organization can pivot quickly in response to new technological developments, reallocate resources effectively, and embrace new approaches without being encumbered by rigid structures or outdated processes. This requires a leadership style that encourages experimentation, tolerates failure as a learning opportunity, and is open to reimagining business models and operational frameworks. The ability to rapidly adopt and integrate new AI capabilities will be a critical determinant of competitive advantage in the coming years.

Ultimately, anticipating the next wave of AI advancements is not about possessing a crystal ball, but about cultivating a strategic discipline. It involves a commitment to continuous learning, openness to new ideas, and a willingness to engage with the complex, rapidly evolving world of artificial intelligence. Leaders who actively build this foresight will be better equipped to seize emerging opportunities, mitigate potential threats, and guide their organizations and teams toward a future where AI is a powerful enabler of innovation and growth. This proactive stance ensures that leadership remains relevant and effective, not just in the present AI era, but in the many AI-driven eras yet to come.

As we have established, the relentless pace of artificial intelligence development necessitates a fundamental re-evaluation of leadership attributes. While foresight and strategic anticipation are crucial for navigating AI's evolution, it is the twin pillars of adaptability and resilience that equip leaders to manage the journey effectively amid its transformative impact. In this dynamic environment, the ability to pivot, to absorb shocks, and to maintain forward momentum in the face of unforeseen disruptions is no longer a desirable trait; it is an essential survival mechanism for both leaders and their organizations.

Adaptability in the AI era transcends mere flexibility. It is the proactive and intentional capacity to adjust strategies, processes, and even organizational structures in response to emerging AI capabilities, market shifts, and evolving competitive landscapes. This involves a conscious effort to remain agile, to embrace new tools and methodologies, and to foster an environment where learning and unlearning are continuous activities.

For leaders, this begins with cultivating a personal disposition towards change. It requires a deliberate shedding of rigid thinking patterns and a genuine embrace of the unknown. This might manifest as a willingness to experiment with novel AI-powered tools, even if their immediate ROI is uncertain, or to redesign workflows that have long been considered immutable. It is about understanding that what worked yesterday might be obsolete tomorrow, and actively seeking ways to be at the forefront of that change rather than a passive recipient.

This can involve dedicating time for personal learning, specifically on how AI is reshaping decision-making, operational efficiencies, and customer engagement within your industry and beyond. Leaders might subscribe to newsletters focused on AI implementation case studies, participate in workshops on agile methodologies tailored for AI integration, or engage consultants specializing in technology-driven organizational transformation. The key is a proactive, rather than reactive, stance towards learning and unlearning.

Beyond personal disposition, fostering organizational adaptability is paramount. This involves building systems and cultures that are inherently more responsive. One critical aspect is fostering cross-functional collaboration. AI often impacts multiple departments simultaneously. When silos are broken down and teams are encouraged to share insights and collaborate on AI integration projects, the organization becomes more nimble.

Imagine a marketing team discovering a new AI tool for hyper-personalized customer segmentation. Without collaboration, this discovery might remain siloed. However, if they proactively engage with the IT department and the sales team, they can collaboratively explore how this AI tool can inform product development, enhance sales outreach, and create a truly unified customer experience. This requires leaders to champion communication channels, both formal and informal, that encourage such interdisciplinary dialogue. Establishing "AI innovation labs" or "cross-functional AI task forces" can provide structured avenues for this. These units, comprised of individuals from diverse backgrounds—from data scientists and engineers to marketing specialists and operational managers—can then collectively assess, pilot, and integrate new AI solutions. Their varied perspectives ensure that AI adoption is not only technically sound but also strategically aligned with broader business objectives and customer needs.

Furthermore, a commitment to continuous upskilling and reskilling is a cornerstone of organizational adaptability. As AI automates certain tasks and creates new roles, leaders must invest in their workforce. This is not merely about providing generic training; it is about identifying the specific AI-related competencies that will be critical for the organization's future success and developing targeted programs to cultivate them. This might include training in AI literacy for all employees, enabling them to understand AI's capabilities and limitations, and specialized training for technical teams on AI development, deployment, and maintenance.

Leaders must champion a culture where learning is not an occasional event but an ongoing process, integrated into the daily workflow. This can be supported by providing access to online learning platforms, funding attendance at relevant conferences and workshops, and creating internal mentorship programs that pair experienced employees with others to navigate new AI tools and concepts. Recognizing and rewarding individuals and teams who proactively acquire and apply new AI-related skills reinforces this culture of continuous learning.

The second indispensable leadership trait in the AI era is resilience. Resilience is the capacity to withstand, adapt to, and recover from adversity, disruption, and setbacks. In a landscape shaped by the unpredictable nature of AI innovation, this translates to an organization's ability to maintain operational continuity, psychological well-being, and strategic focus even when faced with unexpected challenges, such as the sudden emergence of a disruptive AI competitor, significant data breaches impacting AI models, or the ethical dilemmas posed by AI deployments. Resilient leaders are not immune to stress; instead, they possess effective strategies for managing it and preventing it from derailing their teams and operations.

One fundamental aspect of building resilience is cultivating a robust and forward-thinking operational framework. This involves anticipating potential points of failure within AI systems and developing contingency plans. For instance, if an organization relies heavily on a specific AI-powered predictive maintenance system, resilience planning might involve identifying alternative data sources, establishing manual override procedures, or exploring redundant AI models that can take over if the primary one fails.

This proactive risk management, often referred to as "pre-mortem analysis"—imagining a project has failed and then working backward to identify causes—can be particularly effective in identifying vulnerabilities in AI deployments before they become critical issues. Leaders should encourage their teams to engage in these exercises, fostering a mindset of cautious optimism and preparedness. This proactive approach extends to cybersecurity, as AI systems can become new targets. Robust security protocols, regular audits, and rapid incident response plans are essential to protect both data and operational integrity.

On a more personal and team-oriented level, resilience is deeply intertwined with psychological well-being and a strong sense of purpose. Leaders play a crucial role in creating an environment where individuals feel psychologically safe to voice concerns, admit mistakes, and seek support. The rapid evolution of AI can create anxiety and uncertainty among employees, who may fear job displacement or the inability to keep pace with new technologies. Resilient leaders acknowledge these anxieties and address them openly and empathetically.

This involves clear and consistent communication about the organization's AI strategy, its implications for roles and responsibilities, and the support systems available for employees. For example, during a period of significant AI-driven automation, a leader might hold town hall meetings to explain how jobs will evolve, what reskilling opportunities will be provided, and how the organization plans to redeploy talent. This transparency helps build trust and reduce fear, both critical components of team resilience.

Cultivating a "growth mindset" is another vital strategy for fostering resilience. This mindset emphasizes that abilities and intelligence can be developed through dedication and hard work. In the AI era, this means viewing challenges not as insurmountable obstacles but as opportunities for learning and growth. When an AI project encounters unexpected difficulties, a leader with a growth mindset will encourage their team to analyze the setback, extract lessons learned, and adapt their approach, rather than succumbing to discouragement.

This is often facilitated by reframing failures as learning experiences. Instead of punishing mistakes, leaders can create an environment where it is safe to experiment and where lessons from failures are systematically captured and shared. For instance, after a pilot AI project yields unexpected results, a leader might convene a post-mortem meeting to understand why it did not work, what hypotheses were tested, and what new insights were gained, regardless of the outcome. This approach encourages innovation and learning, building the team's capacity to handle future challenges.

Learning from setbacks is not merely about internal reflection; it also involves embracing feedback and external learning. Resilient leaders are open to constructive criticism, both for themselves and for their strategies. They actively seek out diverse perspectives, understanding that blind spots can be a significant vulnerability, especially in the complex domain of AI. This might involve soliciting feedback from customers, partners, and even competitors, and integrating that input into strategic adjustments. Furthermore, resilient organizations often have mechanisms for continuous improvement built into their processes. This could include regular performance reviews that focus on learning and development, "lessons learned" databases that capture insights from all projects, and structured opportunities for peer review and knowledge sharing. The ability to systematically learn from both successes and failures allows an organization to become progressively more robust and adaptable over time.

Moreover, managing stress and promoting well-being are integral to leader and team resilience. The constant pressure to innovate and adapt in the AI era can lead to burnout. Resilient leaders prioritize their own mental and physical health, recognizing that their capacity to lead effectively is directly tied to their well-being. This involves setting boundaries, delegating effectively, and seeking support when needed. Similarly, they must champion their teams' well-being. This could include encouraging regular breaks, discouraging excessive overtime, promoting work-life balance, and providing access to mental health resources. A burnt-out leader or team is inherently less resilient and less able to navigate the inevitable challenges that AI will present. Initiatives such as mindfulness programs, flexible work arrangements, and leadership coaching can help build a more resilient workforce.

In essence, developing adaptability and resilience in the AI era is about building an organization that is not only technologically capable but also mentally and operationally robust. It is about fostering a culture that embraces change, learns from adversity, and prioritizes the well-being of its people. Leaders who cultivate these traits in themselves and their teams will be better equipped not just to survive but to thrive in the face of AI-driven transformation, ensuring sustained performance and a positive, productive work environment for all. This continuous process of adaptation and the unwavering pursuit of resilience are the bedrock upon which future-proof leadership is built, enabling organizations to navigate the complexities of the AI revolution with confidence and agility.

In the accelerating landscape of artificial intelligence, leadership's role shifts from that of a steadfast captain steering a known course to that of a skilled navigator charting an ever-changing sea. While foresight and strategic anticipation guide AI's evolution, it is the twin pillars of adaptability and resilience that equip leaders to manage the journey effectively amid its transformative impact. This necessitates a fundamental re-evaluation of what it means to lead, moving beyond traditional command-and-control structures to embrace dynamic, responsive, and continually evolving models.

Adaptability in the AI era transcends mere flexibility. It is the proactive and intentional capacity to adjust strategies, processes, and even organizational structures in response to emerging AI capabilities, market shifts, and evolving competitive landscapes. This involves a conscious effort to remain agile, to embrace new tools and methodologies, and to foster an environment where learning and unlearning are continuous activities.

For leaders, this begins with cultivating a personal disposition towards change. It requires a deliberate shedding of rigid thinking patterns and a genuine embrace of the unknown. This might manifest as a willingness to experiment with novel AI-powered tools, even if their immediate ROI is uncertain, or to redesign workflows that have long been considered immutable. It is about understanding that what worked yesterday might be obsolete tomorrow, and actively seeking ways to be at the forefront of that change rather than a passive recipient. This can involve dedicating time for personal learning, specifically on how AI is reshaping decision-making, operational efficiencies, and customer engagement within your industry and beyond. Leaders might subscribe to newsletters focused on AI implementation case studies, participate in workshops on agile methodologies adapted for AI integration, or engage consultants who specialize in technology-driven organizational transformation. The key is a proactive, rather than reactive, stance towards learning and unlearning.

Beyond personal disposition, fostering organizational adaptability is paramount. This involves building systems and cultures that are inherently more responsive. One critical aspect is fostering cross-functional collaboration. AI often impacts multiple departments simultaneously. When silos are broken down and teams are encouraged to share insights and collaborate on AI integration projects, the organization becomes more nimble. Imagine a marketing team discovering a new AI tool for hyper-personalized customer segmentation. Without collaboration, this discovery might remain siloed. However, if they proactively engage with the IT department and the sales team, they can collaboratively explore how this AI tool can inform product development, enhance sales outreach, and create a truly unified customer experience.

This requires leaders to champion communication channels, both formal and informal, that encourage such interdisciplinary dialogue. Establishing “AI innovation labs” or “cross-functional AI task forces” can provide structured avenues for this. These units, comprised of individuals from diverse backgrounds—from data scientists and engineers to marketing specialists and operational managers—can then collectively assess, pilot, and integrate new AI solutions. Their diverse perspectives ensure that AI adoption is not only technically sound but also strategically aligned with broader business objectives and customer needs.

Furthermore, a commitment to continuous upskilling and reskilling is a cornerstone of organizational adaptability. As AI automates specific tasks and creates new roles, leaders must invest in their workforce. This isn't merely about providing generic training; it's about identifying the specific AI-related competencies that will be critical for the organization's future success and developing targeted programs to cultivate them. This might include training in AI literacy for all employees, enabling them to understand AI's capabilities and limitations, and specialized training for technical teams on AI development, deployment, and maintenance. Leaders must champion a culture where learning is not an occasional event but an ongoing process, integrated into the daily workflow. This can be supported by providing access to online learning platforms, funding attendance at relevant conferences and workshops, and creating internal mentorship programs that pair experienced employees with others to navigate new AI tools and concepts. Recognizing and rewarding individuals and teams who proactively acquire and apply new AI-related skills reinforces this culture of continuous learning.

The second indispensable leadership trait in the AI era is resilience. Resilience is the capacity to withstand, adapt to, and recover from adversity, disruption, and setbacks. In a landscape shaped by the unpredictable nature of AI innovation, this translates to an organization's ability to maintain operational continuity, psychological well-being, and strategic focus even when faced with unexpected challenges, such as the sudden emergence of a disruptive AI competitor, significant data breaches impacting AI models, or the ethical dilemmas posed by AI deployments. Resilient leaders are not immune to stress; instead, they possess effective strategies for managing it and preventing it from derailing their teams and operations.

One fundamental aspect of building resilience is cultivating a robust and forward-thinking operational framework. This involves anticipating potential points of failure within AI systems and developing contingency plans. For instance, if an organization relies heavily on a specific AI-powered predictive maintenance system, resilience planning might involve identifying alternative data sources, establishing manual override procedures, or exploring redundant AI models that can take over if the primary one fails. This proactive risk management, often referred to as "pre-mortem analysis" (imagining a project has failed and then working backward to identify causes), can be particularly effective in identifying vulnerabilities in AI deployments before they become critical issues. Leaders should encourage their teams to engage in these exercises, fostering a mindset of cautious optimism and preparedness. This proactive approach extends to cybersecurity, as AI systems can become new targets. Robust security protocols, regular audits, and rapid incident response plans are essential to protect both data and operational integrity.

Cultivating a "growth mindset," as popularized by Carol Dweck, is another vital strategy for fostering resilience. This mindset emphasizes that abilities and intelligence can be developed through dedication and hard work. In the AI era, this means viewing challenges not as insurmountable obstacles but as opportunities for learning and growth. When an AI project encounters unexpected difficulties, a leader with a growth mindset will encourage their team to analyze the setback, extract lessons learned, and adapt their approach, rather than succumbing to discouragement. This is often facilitated by reframing failures as learning experiences. Instead of punishing mistakes, leaders can create an environment where it is safe to experiment and where lessons from failures are systematically captured and shared. For instance, after a pilot AI project yields unexpected results, a leader might convene a post-mortem meeting to understand why it did not work, what hypotheses were tested, and what new insights were gained, regardless of the outcome. This approach encourages innovation and learning, building the team's capacity to handle future challenges.

Moreover, the manager's role in talent development will be significantly reshaped. Instead of focusing solely on traditional performance metrics, managers will increasingly assess individuals on their ability to adapt, learn, and collaborate with AI systems. This means recognizing and rewarding those who can effectively leverage

AI tools to enhance their work, identify new applications for AI, and demonstrate strong ethical judgment in its deployment. Performance reviews might incorporate criteria such as the ability to formulate effective AI prompts, interpret AI-generated insights, and integrate AI outputs into complex decision-making processes. This proactive approach to talent management ensures that the workforce remains not only competitive but also aligned with the evolving demands of an AI-driven economy. A forward-thinking manager, for instance, might identify a junior analyst who shows a natural aptitude for interacting with a new AI-powered customer sentiment analysis tool and provide them with advanced training and opportunities to lead its implementation across different departments, thereby fostering both individual growth and organizational advantage.

The concept of "management by exception" will also gain prominence. With AI handling many routine operational tasks and flagging anomalies, managers can dedicate more time to strategic planning, complex problem-solving, and fostering innovation. AI systems can monitor performance metrics, identify deviations from targets, and even suggest corrective actions. The manager's role then becomes to review these exceptions, apply their judgment and experience to decide on the best course of action, and communicate these decisions effectively to the relevant stakeholders. This allows managers to operate at a higher strategic level, focusing their expertise on situations that require human discernment, creativity, and ethical consideration. Imagine an AI monitoring supply chain logistics and flagging a potential disruption due to an unexpected geopolitical event. Instead of spending hours manually sifting through data, the manager can receive an AI-generated alert with proposed mitigation strategies, enabling them to quickly assess the situation, make strategic decisions, and communicate with suppliers and clients, thereby minimizing potential damage.

Furthermore, the manager will increasingly act as an ethical steward in the AI-augmented workplace. As AI systems become more sophisticated, they will inevitably raise ethical dilemmas related to data privacy, algorithmic bias, job displacement, and transparency in decision-making. Managers will be at the forefront of navigating these complex issues, ensuring that AI is deployed responsibly and in alignment with organizational values and societal expectations. This requires not only a strong ethical compass but also the ability to engage in open and transparent communication with their teams and stakeholders about the use of AI. For example, if an AI system used for hiring is found to exhibit bias against certain demographic groups, the manager must be equipped to address this issue proactively, working with AI specialists to rectify the prejudice and communicate the steps taken to the affected parties. This proactive ethical leadership builds trust and ensures that AI is used as a force for good, rather than a source of harm.

The role of inspiration and motivation also takes on new dimensions. In an environment where AI handles many repetitive tasks, managers have a greater opportunity to focus on inspiring their teams, fostering a sense of purpose, and building a strong organizational culture. This might involve articulating a compelling vision for how AI can be used to achieve ambitious goals, celebrating successes from human-AI collaboration, and providing opportunities for employees to engage in meaningful, creative work. The manager acts as a champion for the human element in an increasingly automated world, ensuring that technology empowers rather than disempowers individuals. Consider a scenario where AI automates a significant portion of a customer service team's workload. The manager's task then becomes to reorient the team toward more complex customer issues, proactive customer success initiatives, and building deeper relationships, thereby transforming the role from reactive problem-solving to proactive value creation, all while maintaining a high level of team morale and engagement.

In essence, the evolving role of the manager in an AI-augmented workplace is one of strategic facilitation, ethical stewardship, and cultivation of human capability. It is a role that demands a blend of technical understanding, emotional intelligence, and a forward-looking vision. AI will not replace managers; rather, it will empower them, enabling them to focus on the uniquely human aspects of leadership that are critical for navigating the complexities of the future of work. By embracing AI as a partner, managers can amplify their own effectiveness and that of their teams, ensuring continued relevance and success in an era defined by intelligent technology. This transition is not merely about adopting new tools; it is about fundamentally rethinking how we lead, manage, and collaborate in a world increasingly shaped by artificial intelligence. It is about ensuring that as technology advances, so too does our capacity for insightful, empathetic, and visionary leadership. The manager becomes less of a supervisor and more of a strategic partner, a catalyst for human potential, and a guardian of ethical technological integration. This proactive adaptation is not just about staying relevant; it is about leading the charge into a new era of human-AI synergy, unlocking unprecedented levels of productivity, innovation, and human fulfillment.

As we stand at the precipice of an era undeniably shaped by artificial intelligence, the question of leadership legacy becomes not just a philosophical contemplation but a practical imperative. The decisions we make today, the ethical frameworks we establish, and the very ethos we infuse into our organizations will reverberate far beyond our immediate tenure. To lead effectively in this AI-augmented landscape is to architect a future, and the gravity of that responsibility weighs heavily. It compels us to move beyond the immediate demands of operational efficiency and toward a more profound consideration of the lasting impact of our leadership. What kind of world are we helping to build with the tools we deploy and the principles we uphold? This retrospective lens is crucial for any leader aiming to make a positive and enduring contribution.

Conclusion: The Gravity of Command

The book began with a simple premise: Leadership is the gravity that holds the mission together. As we close, that truth is stronger than ever.

AI will write the code. It will optimize the schedule. It will predict the risks. But AI cannot look a terrified stakeholder in the eye and give them confidence. It cannot mentor a junior developer through imposter syndrome. It cannot choose to do the hard, right thing over the easy, profitable thing.

That is your job. That is your gravity.

Bob turned to Alice. “You know, I think I’m ready to retire. The ship is steady.”

Alice shook her head. “Not yet, Bob. We just installed the quantum computing module. I’m going to need you to explain ‘Stakeholder Management’ to the new physics engine.”

Bob laughed. “Suit up, then. Let’s take command.”

9. Actionable Frameworks for AI Leadership

The Commander's Console

"Theory is good, but checklists save lives." — Bob's Desk Plaque

We have journeyed through the solar system of leadership, from the scorching speed of Mercury to the grounding force of Earth. We have debated ethics, navigated bias, and learned to translate tech-speak into boardroom strategy.

Now, we land.

This final chapter is your Mission Control Console. It strips away the narrative, leaving you with the raw tools. But these aren't just random templates; they are aligned directly with the PMBOK® Guide (7th Edition) Performance Domains. Whether you are managing a waterfall migration or an agile AI sprint, these frameworks ensure your "Gravity" is compliant, structured, and effective.

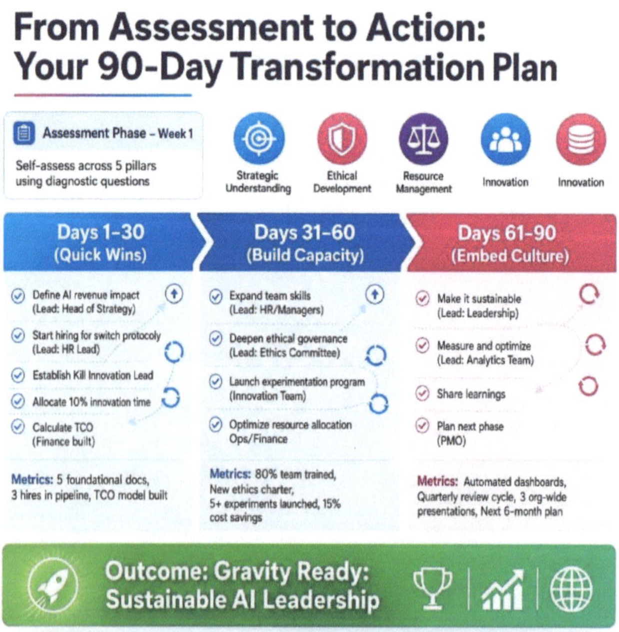

1. The AI Leadership Readiness Assessment

PMBOK Alignment: Team Performance Domain & Leadership Principle

Use this diagnostic to evaluate where you stand today and where you need to grow.

Scoring: Rate yourself 1 (Novice) to 5 (Expert).

Pillar	Assessment Question	Score (1-5)	Action to Improve
Strategic Understanding	Can I articulate exactly how AI drives revenue or cuts costs for *my specific* business unit? [2]		*Read 3 industry case studies this week.*
Team Development	Am I hiring for "adaptability" and "curiosity" over just keywords and tenure? [3]		*Revise one job description today.*
Ethical Governance	Do I have a clear protocol ("Kill Switch") for when an AI model shows bias? [4]		*Draft a 1-page ethics policy.*
Innovation	Do I allocate at least 10% of my team's time to experimenting with tools that might fail? [5]		*Schedule an "Innovation Sprint."*
Resource Management	Can I calculate the Total Cost of Ownership (TCO) for an AI tool, including maintenance and training? [6]		*Audit your current cloud spend.*

1. The AI Leadership Readiness Assessment

PMBOK Alignment: Team Performance Domain & Leadership Principle

The AI Leadership Readiness Assessment is built upon several key pillars, each representing a crucial facet of leading in an AI-augmented world.

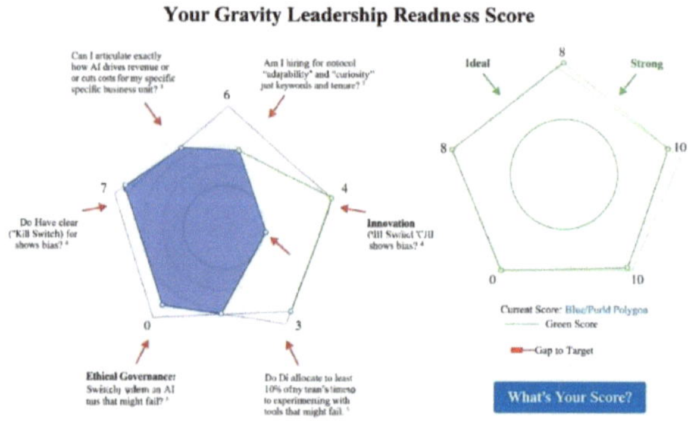

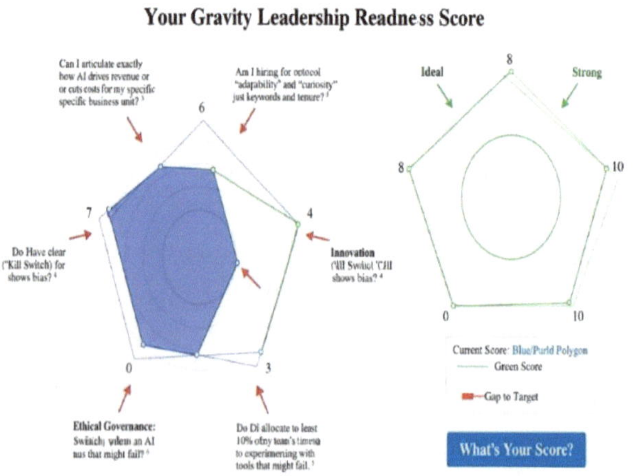

Pillar 1: Strategic Understanding of AI

This dimension probes the manager's grasp of AI's potential business applications, its limitations, and its broader impact on industry and society. It delves into whether the manager can articulate a clear vision for how AI can drive organizational goals, identify relevant AI use cases, and understand the competitive landscape shaped by AI technologies.

A strong strategic understanding means not just recognizing AI as a buzzword, but also comprehending its underlying principles, developmental trajectory, and capacity to disrupt existing business models or create entirely new ones. This includes an awareness of different AI methodologies, such as machine learning, natural language processing, and computer vision, and an appreciation for how these can be applied to solve specific business problems. Furthermore, it involves understanding the data requirements for AI initiatives, the ethical implications of data usage, and the potential return on investment.

Managers are encouraged to consider questions like:

- "Can I clearly explain how AI can benefit our specific business unit or organization?"
- "Do I understand the competitive advantages AI can offer, and how our competitors are leveraging it?"
- "Am I aware of the potential risks and ethical considerations associated with AI adoption within our context?"

Self-scoring within this domain involves rating one's confidence in articulating an AI strategy, identifying opportunities for AI implementation, or assessing the feasibility of AI projects. For instance, a manager might rate their ability to identify a viable AI use case for their department on a scale of 1 to 5. This could be further broken down into specific applications, such as the "Ability to identify opportunities for AI in customer service" or "Understanding of AI's potential for process automation." The objective is to pinpoint knowledge gaps, guiding further learning and strategic planning. This might involve seeking out industry reports, attending AI-focused workshops, or engaging with AI experts to deepen foundational knowledge.

Pillar 2: Team Development and Management Capabilities

Leading AI initiatives is not solely about technology; it is profoundly about the people who develop, deploy, and interact with these systems. This dimension assesses a manager's ability to build and nurture diverse, high-performing teams that can work effectively with AI. It examines skills in recruiting talent with the right technical and soft skills, fostering collaboration between technical experts and domain specialists, and creating an environment that encourages continuous learning and adaptation.

In the context of AI, this means understanding the unique skill sets required, such as data science, machine learning engineering, AI ethics, and human-AI interaction design. It also involves recognizing the importance of soft skills like critical thinking, problem-solving, communication, and ethical reasoning, which become even more vital as AI automates more routine tasks.

The assessment asks:

- "How effectively can I build and manage a team with the necessary AI-related expertise?"
- "Do I foster a culture of collaboration and knowledge sharing among team members, including those with non-technical backgrounds?"
- "Am I adept at identifying skill gaps and providing opportunities for my team to upskill and reskill in response to AI advancements?"

A manager might evaluate their comfort level in discussing AI concepts with technical staff, their approach to performance management for AI-focused roles, or their strategies for integrating AI into existing workflows without alienating current employees. For example: "On a scale of 1 to 5, how confident am I in my ability to mediate discussions between data scientists and business stakeholders to ensure alignment on AI project objectives?"

Further granularity could involve assessing proficiency in identifying and mitigating potential conflicts, such as fear of job displacement or differing technical perspectives. This might lead to a realization that further training in change management or cross-functional team leadership is needed. The emphasis is on developing leadership practices that empower individuals, promote psychological safety, and cultivate a growth mindset. This could involve implementing regular feedback mechanisms, creating dedicated learning time, or facilitating cross-functional idea sharing.

Pillar 3: Ethical Awareness and Governance

Thirdly, "Ethical Awareness and Governance" is paramount. The responsible deployment of AI hinges on a robust understanding of its ethical implications and the establishment of clear governance frameworks. This dimension assesses a manager's comprehension of AI ethics, including issues such as bias, fairness, transparency, accountability, and privacy. It examines their ability to anticipate potential ethical challenges, implement safeguards, and foster a culture of ethical responsibility within their teams.

Leaders must be equipped to grapple with complex questions surrounding algorithmic bias, the potential for AI to perpetuate or exacerbate societal inequalities, and the imperative of data privacy in an era of vast data collection.

The assessment questions here might include:

- "Do I understand the common sources of bias in AI systems and how to mitigate them?"
- "Am I prepared to establish and enforce ethical guidelines for AI development and deployment?"
- "Do I have a clear understanding of data privacy regulations (e.g., GDPR, CCPA) and how they apply to AI projects?"
- "Can I articulate the principles of explainable AI and the importance of transparency?"

A manager might rate their familiarity with AI ethics frameworks or their confidence in identifying and addressing potential biases in a proposed AI model. For example, "How effective am I at conducting an ethical risk assessment for a new AI application?" could be scored. This might reveal a need for managers to educate themselves on specific ethical frameworks, attend workshops on AI bias mitigation, and engage with legal and compliance departments to ensure regulatory compliance. It also involves cultivating an environment where team members feel empowered to raise ethical concerns without fear of reprisal, fostering a proactive approach to ethical AI development. This could include establishing an AI ethics committee, incorporating ethical review stages into project lifecycles, or providing training on responsible AI practices for all stakeholders. The goal is to embed ethical considerations as a core component of AI strategy and execution, not as an afterthought.

Pillar 4: Innovation and Adaptability

The fourth pillar is Innovation and Adaptability. The AI landscape is characterized by rapid evolution, making a leader's capacity for innovation and adaptability crucial for sustained success. This dimension assesses a manager's openness to experimentation, their willingness to embrace new technologies and methodologies, and their agility in responding to change. It looks at how well they foster a culture that encourages calculated risk-taking, learns from failures, and remains flexible in the face of unforeseen challenges.

Leading AI innovation requires not only embracing cutting-edge tools but also understanding how to integrate them into existing processes and how to pivot when strategies are no longer effective.

The assessment probes:

- "How comfortable am I with piloting new AI technologies and methodologies?"
- "Do I actively encourage experimentation and learning from both successes and failures within my team?"

- "Am I capable of quickly adapting project plans and strategies in response to new AI developments or changing market conditions?"
- "Do I foster an environment where creativity and novel solutions are valued and explored?"

A manager might assess their track record in championing novel AI approaches or their response to unexpected project roadblocks. For instance, "To what extent do I proactively seek out emerging AI trends and evaluate their potential relevance to our business?" could be a self-evaluation point. A low score here might indicate a need for managers to dedicate time to staying abreast of AI research, allocate resources for R&D or pilot projects, and develop more agile project management methodologies. It is about cultivating a mindset that views change not as a threat, but as an opportunity for growth and improvement, ensuring the organization remains at the forefront of AI-driven innovation. This could involve establishing innovation labs, implementing agile development sprints for AI projects, or encouraging cross-functional "hackathons" to explore new AI applications.

Pillar 5: Resource Management and Investment Acumen

Finally, Resource Management and Investment Acumen form the fifth pillar. Implementing AI solutions often requires significant investment in technology, talent, and infrastructure. This dimension evaluates a manager's ability to identify, justify, and effectively manage the resources needed for AI initiatives, ensuring a strong return on investment. It covers their skills in budgeting, project prioritization, and understanding the financial implications of AI adoption, including both upfront costs and ongoing operational expenses.

It is about making informed decisions regarding where and how to allocate financial and human capital to maximize the impact of AI.

Key questions include:

- "Can I effectively build a business case for AI investments, clearly articulating the expected ROI?"
- "Do I understand the total cost of ownership for AI solutions, including infrastructure, maintenance, and talent?"
- "How proficient am I in prioritizing AI projects based on strategic alignment and potential impact?"
- "Am I able to secure the necessary funding and resources for AI initiatives?"

A manager might assess their success in obtaining budget approval for a previous AI project or their ability to forecast the ongoing operational costs of an AI system. For example, "How well do I understand the interplay between AI investment and long-term business strategy?" could be a crucial point for self-reflection. A lower score might signal a need for training in financial modeling, business case development, or strategic investment analysis, particularly in relation to technology adoption. It emphasizes the need for leaders to be not just technologically savvy but also financially astute stewards of the organization's resources when pursuing AI-driven transformations. This could involve collaborating closely with finance departments, developing robust metrics for measuring AI ROI, and conducting thorough due diligence on potential AI vendors and solutions.

2. The "Mars Rover" Project Checklist

PMBOK Alignment: Development Approach and Life Cycle & Uncertainty Performance Domain (Risk)

Use this before launching any AI pilot to prevent a "Titanic Moment" by managing complexity and risk.

Phase 1: The Pre-Flight Check (Initiation)

- **[ ] The "Why" Test:** Can every member of the team explain the business value of this pilot in one sentence? (Value Delivery).

- **[ ] Data Hygiene:** Has the training data been screened for historical bias? (Quality Management).
- **[ ] The "Human in the Loop":** Have we identified which high-stakes decisions require human review before the AI acts? (Stewardship).

Phase 2: The Orbit Check (Execution)

- **[ ] The Drudgery Audit:** Is the team using AI to automate their own administrative tasks, such as status reports or meeting notes? (Resource Optimization).
- **[ ] The Silence Check:** If the metrics look too good, is anyone speaking up? (Stakeholder Engagement).
- **[ ] The Feedback Loop:** Is there a simple channel for end-users to report "This AI is confusing" without filling out a formal ticket? (Adaptability).

*****

3. The Stakeholder Communication Matrix

PMBOK Alignment: Stakeholder Performance Domain

Use this to tailor your message so you do not bore the Board with Python code or scare the engineers with marketing fluff.

Who	Care Abouts	The "Dialect" to Use	Frequency
Executive Board	ROI, Risk, Reputation	**Strategic:** "This tool saves 5,000 hours." (Not "F1-Score is .92") [14]	Monthly
Technical Team	Feasibility, Specs, Tech Debt	**Precise:** "The API latency is 200ms." (Not "It's fast") [15]	Daily
End-Users	Workflow, "Will I lose my job?"	**Empathetic:** "This automates the boring data	Weekly

Who	Care Abouts	The "Dialect" to Use	Frequency
		entry so you can do analysis." [16]	
Legal/Compliance	Liability, Data Privacy	**Risk-Averse:** "Here is our bias mitigation log and audit trail." [17]	Quarterly

4. The AI Business Case Builder

PMBOK Alignment: Value Performance Domain (Benefits Realization)

Use this framework to get budget approval.

1. The Hook (The Problem):

"Currently, our [Team Name] spends [X] hours per week on [Manual Task]. This costs us $[Y] annually and causes [Z]% employee churn."

2. The Solution (The AI):

"We propose implementing [AI Tool] to automate [Specific Task]. This is not a replacement for staff; it is an augmentation."

3. **The ROI (The Math):**
   - **Hard Savings:** $[Amount] in reduced operational costs/overtime.
   - **Soft Savings:** [Number] hours returned to strategic work.
   - **Risk Reduction:** [Percentage] decrease in human error rate.
4. The Ask (The Investment):

"We need $[Budget] for a 3-month pilot. Success will be measured by [Specific KPI, e.g., adoption rate or time saved]."

*****

5. The "Odd" Leader's Manifesto

PMBOK Alignment: Stewardship Principle (Integrity, Care, Trustworthiness)

A reminder to stay human in a machine world.

1. **Be Oddly Human:** When the bot writes the email, you pick up the phone.
2. **Be Oddly Slow:** When the algorithm says "Go," you pause to ask "Why?"
3. **Be Oddly Curious:** When you feel safe, you disrupt yourself.

*****

AI Project Communication Plan Template

Project Name: [Insert AI Project Name Here]

Date Prepared: [Insert Date]

Prepared By: [Insert Name/Role]

Version: [Insert Version Number]

1. Executive Summary

Briefly outline the purpose of the AI project, its key objectives, and the overarching communication strategy. Emphasize the importance of stakeholder engagement for project success.

2. Stakeholder Identification and Analysis

Identify Key Stakeholders: List all individuals, groups, or departments who have an interest in, influence over, or are affected by the AI project.

- **Example Stakeholders:** Executive Sponsors, Project Team (Data Scientists, Engineers, Project Managers), Business Unit Leaders, End-Users, IT Operations, Legal/Compliance, Marketing, Customers.

Analyze Stakeholder Needs and Interests: For each stakeholder, document their primary interest, level of influence (High/Medium/Low), potential impact, preferred communication methods/frequency, and potential resistance points.

Stakeholder Matrix (Example):

Stakeholder Group	Interest	Influence	Impact	Communication Needs	Preferred Channels	Frequency	Potential Concerns
Executive Sponsors	ROI, strategic alignment, risk mitigation	High	High	High-level progress, key decisions, and budget status	Executive Briefings, PMO Reports	Monthly	Delays, cost overruns
Project Team	Technical details, task clarity, and collaboration	Medium	High	Daily updates, impediment resolution, and design reviews	Daily Stand-ups, Slack, Git Commits	Daily/ Weekly	Scope creep, technical debt
Business Leaders	Functional benefits, user adoption, ROI	High	Medium	Demo of capabilities, impact on operations, progress	Sprint Reviews, Demos, Email Updates	Bi-weekly	Disruption, unproven tech
End-Users	Usability, impact on daily tasks, training	Low	High	How it works, benefits, training sessions, and feedback	Workshops, User Guides, Surveys	As needed	Learning curve, job security
Legal/Compliance	Data privacy, ethical guidelines, and regulation	Medium	High	Data handling, algorithmic bias, audit trails	Dedicated Consultations, Policy Reviews	Quarterly	Non-compliance, reputational risk

3. Communication Objectives

Clearly define what each communication activity aims to achieve.

- To ensure executive sponsors are informed of project progress and risks to secure continued support.
- To gather feedback from business unit leaders on the utility of new AI features.
- To educate end-users on the capabilities and benefits of the AI solution.
- To confirm understanding and adherence to ethical guidelines with the legal team.
- To foster a sense of shared ownership and collaboration within the project team.

4. Communication Channels and Methods

- **Regular Project Updates:** Email newsletters, internal project dashboards, and status reports.
- **Progress Reviews:** Sprint review meetings, demo sessions, and steering committee meetings.
- **Information Sharing:** Project wiki, shared drives, internal collaboration platforms (e.g., Teams, Slack).
- **Feedback Collection:** Surveys, focus groups, user testing sessions, dedicated feedback channels.
- **Escalation:** Defined escalation paths for critical issues and decisions.
- **Formal Documentation:** Project charter, design documents, risk registers, meeting minutes.

5. Communication Schedule and Frequency

- **Daily:** Project team stand-ups.
- **Weekly:** Status reports to key stakeholders, backlog refinement sessions.
- **Bi-weekly:** Sprint review meetings, demos for business unit leaders.

- **Monthly:** Steering committee meetings, executive briefings.
- **Quarterly:** Major project milestone reviews, strategy alignment sessions.
- **Ad Hoc:** Urgent issue resolution, critical decision-making meetings.

6. Key Messages

Define core messages that must be consistently communicated across different stakeholder groups, tailored to their specific interests.

Example Core Message: "Our AI project is progressing iteratively, focusing on delivering measurable business value by [specific metric] within the next quarter. We are actively managing risks related to data bias and operational integration through [specific actions]."

- **For Executives:** "Project on track to achieve [strategic goal]. Current ROI projection is [X]% with a forecasted completion in [Date]. Key upcoming decision: [Decision Point]."
- **For End-Users:** "This new AI tool will streamline your daily tasks by [specific benefit], allowing you to focus on more strategic work. Training sessions will be scheduled starting [Date]."
- **For Legal/Compliance:** "We have implemented [specific safeguards] to ensure data privacy and mitigate algorithmic bias, in line with [relevant regulations]."

7. Roles and Responsibilities

Clearly assign ownership for communication activities to ensure accountability.

- **Project Manager:** Overall responsibility for the Communication Plan, status reporting, and stakeholder coordination.
- **Product Owner:** Communicating the product vision, prioritizing the backlog, and conducting sprint reviews.

- **Technical Lead:** Communicating technical progress, risks, and architectural decisions.
- **Team Members:** Participating in stand-ups and providing updates on task completion.
- **Executive Sponsor:** Communicating project importance, making key decisions, and championing the project across the organization.

8. Feedback Mechanisms

Outline how feedback will be solicited, collected, analyzed, and acted upon.

- Dedicated feedback forms on the project intranet.
- Post-demo Q&A sessions.
- Regular surveys targeting specific user groups.
- A "suggestion box" feature is integrated within the AI application.
- One-on-one meetings with key influencers.

9. Escalation Process

Define how critical issues or concerns will be escalated and resolved.

Example: "The Project Manager will escalate issues impacting critical path timelines or significant budget deviations to the Steering Committee within 24 hours. Decisions will be documented and communicated back to relevant parties within 48 hours."

10. Success Metrics for Communication

How will the effectiveness of the communication plan be measured?

- Stakeholder satisfaction scores (via surveys).
- Level of engagement in feedback sessions.
- Timeliness and completeness of information dissemination.
- Reduction in project-related misunderstandings or conflicts.

- Successful adoption rates of the AI solution.

This structured approach ensures that no stakeholder is left in the dark and that their input is actively sought and considered. It transforms communication from a reactive necessity into a proactive, strategic asset.

*****

Beyond the structured plan, effective AI leadership hinges on managing expectations, especially given the often speculative nature of AI development. Misaligned expectations are a primary source of disappointment and conflict. To counter this, AI leaders must be adept at communicating both the potential and the limitations, the progress and the uncertainties, with clarity and honesty.

Example Scripts for Managing Expectations

Scenario 1: Communicating early-stage uncertainty about model performance.

To Business Stakeholders: "We're currently in the exploratory phase of developing our AI-powered [feature name]. Our initial experiments show promising directions, with early models achieving [e.g., 70%] accuracy on a subset of data. However, achieving the desired [e.g., 95%] accuracy for production use requires further refinement, including [mention specific challenges like data augmentation, hyperparameter tuning, or exploring alternative algorithms]. We're on track to have a clearer picture of the achievable performance ceiling and the timeline for reaching it by the end of the next sprint. Our focus in the coming weeks will be on [specific experimental goal]."

Why it works: It acknowledges progress, quantifies current performance, clearly states the gap, identifies next steps, and provides a timeline for a more definitive update. It avoids over-promising while demonstrating focused effort.

Scenario 2: Explaining delays due to unforeseen data challenges.

To Executive Sponsors: "We've encountered an unexpected challenge with the data pipeline for the [AI project name]. The complexity of integrating and cleaning the historical data from [source name] has proven more significant than initially scoped. This is impacting our timeline for delivering the initial prototype by [original date]. We've re-allocated resources and are working closely with the data engineering team to resolve this. Our revised target for prototype delivery is now [new date]. We are implementing a [new data validation process] to prevent similar issues moving forward. I'm available to discuss the technical details and our mitigation strategy further at your convenience."

Why it works: It takes ownership, clearly states the problem and its impact, provides a revised timeline, outlines corrective actions, and offers further transparency.

Scenario 3: Managing user concerns about job displacement due to automation.

To End-Users: "We understand that introducing new AI tools can raise questions about how it will affect your roles. Our goal with the [AI tool name] is not to replace human expertise, but to augment it. This tool is designed to automate repetitive tasks like [example task], freeing up your time to focus on more complex problem-solving, strategic thinking, and direct customer interaction—areas where human judgment and empathy are crucial. We will provide comprehensive training sessions starting on [Date] to ensure you are comfortable and proficient with the new system. Your feedback throughout this process will be invaluable in shaping its ongoing development and ensuring it truly supports your work."

Why it works: Directly addresses the concern, reframes the AI's purpose as augmentation rather than replacement, highlights the value of human skills, offers concrete support (training), and invites ongoing feedback.

Scenario 4: Communicating ethical considerations and mitigation efforts.

To Legal and Compliance Teams: "As we develop the [AI model name] for [purpose], we are giving paramount importance to ethical considerations and compliance. We have established a dedicated AI ethics review board and are implementing bias-detection and mitigation techniques during data preprocessing and model training. For instance, we are currently analyzing the model's performance across different demographic groups to ensure fairness, and have documented our data provenance and model decision-making processes to ensure transparency and auditability. We would like to schedule a meeting to walk you through our current ethical framework and risk assessment for this project."

Why it works: Proactively addresses ethical concerns, demonstrates a structured approach to mitigation, provides concrete examples of actions taken, and invites collaboration with relevant experts.

*****

These examples highlight a consistent theme: **transparency, honesty, and a focus on solutions.** AI leaders must be prepared to communicate the realities of AI development—the experimentation, the iterative nature, the potential setbacks, and the continuous learning—with the same diligence they apply to managing technical aspects.

To further reinforce stakeholder involvement and ensure no critical aspect is overlooked, the following checklist serves as a self-assessment guide for ensuring consistent and effective engagement.

Stakeholder Engagement and Communication Checklist

I. Planning and Strategy

- [] Have all key stakeholders been identified and their interests analyzed?

- [] Is there a documented communication plan outlining objectives, channels, frequency, and responsibilities?
- [] Are communication objectives clearly defined and measurable?
- [] Are key messages tailored to different stakeholder groups?
- [] Has the communication plan been shared with and validated by key stakeholders (e.g., executive sponsors, project steering committee)?
- [] Are feedback mechanisms clearly defined and accessible to stakeholders?
- [] Is there a clear escalation path for issues and concerns?

II. Execution and Cadence

- [] Are regular project updates being provided as per the communication plan?
- [] Are sprint reviews/demos being conducted effectively to showcase progress and gather feedback?
- [] Are daily stand-ups facilitating intra-team communication and impediment resolution?
- [] Is information about project status, risks, and decisions being disseminated promptly?
- [] Are feedback loops actively managed—is feedback being collected, analyzed, and acted upon?
- [] Are stakeholders being informed of significant changes to scope, timeline, or budget proactively?
- [] Are training and support materials being developed and disseminated for end-users?

III. Content and Clarity

- [] Is technical jargon being translated into business-understandable language when communicating with non-technical stakeholders?
- [] Are the benefits and value proposition of the AI solution clearly articulated?

- [] Are limitations, risks, and uncertainties being communicated honestly and proactively?
- [] Are AI ethics, fairness, and privacy considerations being addressed in communications where relevant?
- [] Are key decisions and their rationales being documented and communicated?
- [] Is progress being measured and communicated against defined KPIs and objectives?

IV. Relationship Management

- [] Are relationships with key stakeholders being actively nurtured?
- [] Are potential conflicts or resistance points being identified and addressed early?
- [] Is there evidence of stakeholder buy-in and support for the project?
- [] Are formal communication records (meeting minutes, reports) being maintained?
- [] Is the communication approach being reviewed and adapted based on stakeholder feedback and project evolution?

*****

This checklist serves as a practical tool for AI leaders to ensure that their stakeholder engagement and communication strategies are not just an afterthought but an integral, well-managed component of the AI leadership framework. It encourages a proactive, systematic, and stakeholder-centric approach, vital for navigating the complexities and maximizing the potential of artificial intelligence initiatives. By consistently applying these tools and principles, AI leaders can build trust, foster collaboration, and ultimately drive the successful adoption and impact of AI within their organizations.

The journey toward becoming an exceptional AI leader is not a static destination but a continuous process of learning, adaptation, and deliberate growth. Throughout this book, we have explored the multifaceted landscape of AI leadership, from understanding the core principles of artificial intelligence and its strategic implications to mastering the art of stakeholder engagement and ethical governance. Now, the critical juncture arrives: translating this acquired knowledge into a concrete, actionable roadmap for your personal and professional evolution. This subsection is dedicated to the creation of your personalized AI leadership development plan—a document that will serve as your compass as you navigate the dynamic and increasingly complex world of AI.

Drawing upon the insights gleaned from the preceding chapters and the self-assessment you have undertaken regarding your current AI leadership readiness, the next step is to define clear, ambitious, yet attainable objectives for your ongoing development. This plan is not a generic template; it is a bespoke blueprint designed to address your unique strengths, acknowledge areas for improvement, and align with your overarching career aspirations within the AI domain.

The fundamental principle guiding this exercise is the establishment of SMART goals: Specific, Measurable, Achievable, Relevant, and Time-bound. By adhering to this framework, you transform abstract intentions into tangible milestones, creating a quantifiable path toward enhanced leadership capabilities.

Reflecting on Your Readiness

Begin by reflecting on your readiness assessment. Consider the key competencies we have discussed: technical acumen, strategic thinking, ethical reasoning, change management, and stakeholder communication.

- Where do you currently stand in each of these areas?
- What are your perceived strengths that can be further honed?
- What are the gaps that require focused attention?

For instance, if your assessment indicated a firm grasp of AI technologies but a need for improvement in communicating complex AI concepts to non-technical executives, this becomes a primary focus area. Alternatively, if you excel at strategic vision but struggle with the practicalities of managing AI project teams through iterative development cycles, that too would shape your objectives.

Formulating SMART Goals

With these reflections in mind, the task is to formulate SMART goals. For each identified area of development, craft a goal that is:

- **Specific:** Clearly define what you aim to achieve. Instead of "improve AI communication," a particular goal would be: "Enhance my ability to articulate the business value and ethical considerations of a new AI initiative to the executive board within 30-minute presentation slots."
- **Measurable:** Establish criteria for success. How will you know when you have achieved your goal? For the communication example, measurability could be defined by positive feedback from the executive board, successful approval of the initiative following your presentation, or a reduction in clarifying questions after the presentation. Other measurable outcomes might include completing a specific number of AI-related training modules or successfully leading an AI pilot project.
- **Achievable:** Ensure your goals are realistic given your current resources, time constraints, and organizational context. While ambition is crucial, setting unattainable goals can lead to frustration. If you aim to become an expert in deep learning overnight, it is unlikely to be achievable; however, completing an introductory online deep learning course within three months is achievable.
- **Relevant:** Your goals must align with your role, your organization's strategic objectives, and your personal career trajectory. If your organization prioritizes the ethical deployment of AI, goals focused on AI ethics and governance will be highly relevant.

- **Time-bound:** Set a clear deadline for achieving each goal. This creates a sense of urgency and provides a framework for tracking progress. Goals can span different time horizons: short-term (1–3 months), medium-term (6–12 months), and long-term (1–3 years). For example: "Complete an online certification in Responsible AI within six months."

Let's illustrate this with a hypothetical AI leader. Suppose their readiness assessment revealed strong technical skills but a recognized weakness in translating AI's potential impact to business stakeholders and a need for greater understanding of AI's ethical implications. Their development plan might then include goals such as:

Goal 1: Enhance Business Communication of AI Value

- **Specific:** Develop and deliver clear, concise, and compelling presentations that articulate the business benefits, ROI, and strategic alignment of AI initiatives to non-technical executive audiences.
- **Measurable:** Receive positive qualitative feedback from at least three executive stakeholders regarding the clarity and persuasiveness of AI project presentations. Successfully secure approval or significant buy-in for at least one AI initiative proposal based on such presentations within the next twelve months.
- **Achievable:** This can be achieved through focused practice, seeking mentorship from experienced business communicators, and using frameworks for translate technical concepts.
- **Relevant:** Directly addresses an identified gap and is crucial for securing resources and organizational support for AI projects.
- **Time-bound:** Within twelve months.

To achieve this, specific actions might include: enrolling in a business communication workshop, seeking opportunities to present at cross-functional meetings, requesting feedback from peers and mentors after presentations, and studying successful AI business case examples.

Goal 2: Deepen Understanding and Application of AI Ethics

- **Specific:** Gain a comprehensive understanding of AI ethics principles, including fairness, accountability, transparency, and privacy, and develop a framework for assessing and mitigating ethical risks in AI deployments within the organization.
- **Measurable:** Complete a recognized online course or professional development program on AI ethics. Develop and propose a draft AI ethics review process for new AI projects, which the relevant governance committee then adopts. Successfully identify and propose mitigation strategies for at least two potential ethical risks in ongoing or planned AI projects.
- **Achievable:** Numerous reputable resources are available, and developing internal processes is feasible with dedicated effort and collaboration.
- **Relevant:** Essential for responsible AI leadership and mitigating reputational and regulatory risks, aligning with a growing emphasis on ethical AI.
- **Time-bound:** Complete the course within six months, and develop and propose the review process within twelve months.

Actionable steps for this goal could include: enrolling in courses offered by leading universities or professional organizations, reading key literature on AI ethics, engaging with legal and compliance teams, and participating in industry forums discussing AI ethics.

Goal 3: Master Agile Methodologies for AI Project Management

- **Specific:** Improve proficiency in applying agile and iterative development methodologies (e.g., Scrum, Kanban) to AI projects, focusing on managing the inherent uncertainties of

AI development, such as model performance variability and data challenges.

- **Measurable:** Successfully led an AI pilot project using an agile framework, completing at least three iterations with demonstrable progress and stakeholder feedback incorporated. Obtain an agile certification (e.g., Certified ScrumMaster) within nine months.
- **Achievable:** Agile principles are adaptable to AI, and certifications are widely available. Leading a pilot project is a practical way to gain hands-on experience.
- **Relevant:** AI development is often best managed through iterative cycles due to its experimental nature, making agile methodologies highly relevant.
- **Time-bound:** Within nine months for certification and successful pilot project completion.

Supporting actions might include: attending agile training, reading books and articles on agile AI development, actively participating in agile ceremonies, and seeking mentorship from experienced agile project managers.

This structured approach to goal setting is the cornerstone of your personalized AI leadership development plan. It provides clarity, direction, and a mechanism for accountability. However, a plan is only as effective as its execution. Therefore, it is crucial to embed these goals within a consistent practice of reflection and action.

Consider establishing a regular cadence for reviewing your progress against these goals—perhaps monthly or quarterly. This review should assess what has been accomplished, what obstacles have been encountered, and what adjustments might be necessary to your plan or actions. This iterative planning, acting, and reviewing process aligns with agile development principles and is essential for navigating the complexities of leadership development.

Furthermore, your development plan should not exist in isolation. Seek opportunities to apply the skills and knowledge you are acquiring in real-world scenarios within your organization. Volunteer for projects that will challenge you in your target areas, seek mentorship from leaders you admire, and actively solicit feedback from your team, peers, and superiors. This practical application is where authentic learning and development occur, transforming theoretical knowledge into lived experience and honed expertise.

*****

Final Transmission

You now have the map. You have the tools. You have the crew. The gravity is yours to generate.

Bob: "Alright, team. Checklists are green. Systems are nominal."
Alice: "Trajectory is locked. We are ready for lift-off."

Execute.

[End of Book]

Back Matter

This appendix serves as a supplementary resource to enhance your practical application of the principles discussed throughout this book. It includes a curated collection of templates, checklists, and frameworks designed to support your AI leadership journey.

- **AI Leadership Readiness Assessment Checklist:** A detailed checklist to guide your self-assessment of key leadership competencies in the AI domain, building upon the self-reflection exercises in the preceding chapters.
- **SMART Goal Setting Worksheet:** A structured worksheet to assist in formulating and documenting your personalized AI leadership development plan, ensuring each objective is Specific, Measurable, Achievable, Relevant, and Time-bound.
- **Stakeholder Engagement Matrix Template:** A practical tool for mapping out key stakeholders, their interests, influence levels, and communication strategies for AI initiatives.
- **AI Ethics Review Framework Outline:** A foundational template for developing an ethical review process for AI projects, covering aspects such as bias detection, transparency, and accountability.
- **Recommended Reading List:** An expanded list of seminal books, influential articles, and reputable online resources for continued learning in AI strategy, ethics, and leadership.

Glossary of Terms

This glossary provides definitions for key terms and concepts introduced in this book. It is intended to serve as a quick reference for understanding the specialized language of artificial intelligence and its application in leadership contexts.

Adaptive Intelligence: The capability of systems—or leaders—to adjust decisions dynamically based on new data, environmental shifts, and evolving constraints.

AI-Augmented Leadership: A leadership approach that combines human judgment with artificial intelligence to enhance foresight, speed, and decision quality without surrendering accountability.

AI Governance: The framework of policies, processes, and responsibilities established to ensure that AI systems are developed and deployed responsibly, ethically, and in alignment with organizational values and regulations.

Algorithmic Drift: The gradual misalignment between an AI system's outputs and the organization's strategic intent due to data shifts, bias, or lack of recalibration.

Artificial Intelligence (AI): The simulation of human intelligence processes by machines, including learning, problem-solving, reasoning, and decision-making.

Bias (in AI): Systematic and repeatable errors in an AI system that result in unfair or skewed outcomes, often stemming from biased training data or flawed algorithm design.

Black Hole Work: Repetitive, low-value tasks that consume time and energy without creating proportional strategic impact; often prime candidates for automation.

Cognitive Load Threshold: The point at which a leader or team becomes overwhelmed by complexity, reducing clarity, speed, and decision quality.

Command Horizon: The visible boundary of influence and accountability a leader can realistically maintain at a given time.

Data Gravity: The tendency of large datasets to attract applications, services, and decision processes toward them, shaping organizational structure and workflows.

Decision Velocity: The speed at which decisions are made and executed without sacrificing strategic coherence or accountability.

Deep Learning (DL): A subset of machine learning that uses multi-layered artificial neural networks to analyze and learn from data, particularly for complex tasks such as image and speech recognition.

Ethical AI: AI systems designed and deployed with principles of fairness, accountability, transparency, privacy, and safety at their core.

Ethical Orbit: The moral and governance boundaries within which AI and leadership decisions must operate.

Explainable AI (XAI): Methods and techniques that allow human users to understand, interpret, and trust the outputs produced by AI systems.

Force Multiplier: A capability—often AI, automation, or strong team alignment—that amplifies leadership effectiveness beyond linear effort.

Friction Cost: Hidden operational drag caused by unclear priorities, duplicated work, misaligned systems, or inefficient processes.

Governance Spine: The structured backbone of policies, controls, and accountability mechanisms that stabilize high-speed digital operations.

Human-in-the-Loop: A design principle ensuring that human oversight remains integrated into automated or AI-driven processes.

Leadership Inertia: Resistance to change caused by legacy thinking, outdated structures, or fear of technological disruption.

Machine Learning (ML): A subset of artificial intelligence that enables systems to learn from data and improve performance without being explicitly programmed.

Mission Drift: Gradual deviation from an organization's core purpose due to reactive decisions, misaligned incentives, or strategic distraction.

Natural Language Processing (NLP): A field of AI focused on enabling computers to understand, interpret, and generate human language.

Organizational Orbit: The predictable patterns of communication, decision-making, and authority that define how a team or organization operates.

Predictive Command: The use of AI-driven forecasting and analytics to anticipate risks, resource needs, and performance outcomes before disruption occurs.

Psychological Gravity: The emotional and cultural force that keeps teams aligned, stable, and resilient under pressure.

Responsible Autonomy: Delegating decision authority—whether to teams or AI systems—while retaining ultimate accountability.

ROI (Return on Investment): A performance measure used to evaluate the efficiency or profitability of an investment relative to its cost.

Signal-to-Noise Ratio: The balance between meaningful insight and distracting or irrelevant data within decision environments.

Stakeholder Management: The systematic process of identifying, analyzing, and engaging individuals or groups who have an interest in or are affected by an initiative.

Strategic Mass: The accumulated influence of credibility, trust, and capability that allows a leader's direction to carry weight within an organization.

Systemic Resilience: An organization's ability to absorb shocks, adapt quickly, and recover without losing strategic direction.

Trust Bandwidth: The amount of complexity, ambiguity, or change a team can handle before trust begins to erode.

Velocity without Stability: High operational speed without adequate governance or control, often leading to burnout, errors, or ethical risks.

Zero-Drift Leadership: A disciplined approach ensuring that rapid innovation and acceleration do not compromise mission clarity, accountability, or ethical standards.

Key References and Recommended Reading

The insights presented in this book are built upon a vast foundation of academic research, industry reports, and expert commentary. The following are key references and recommended readings that have significantly shaped the understanding of AI leadership and its strategic implementation.

- Brynjolfsson, E., & McAfee, A. (2014). The second machine age: Work, progress, and prosperity in a time of brilliant technologies. W. W. Norton & Company.
- Domingos, P. (2015). The master algorithm: How the quest for the ultimate learning machine will remake our world. Basic Books.
- Floridi, L., & Cowls, J. (2019). A unified framework of five principles for AI in society. Harvard Data Science Review, 1(1).
- Goodfellow, I., Bengio, Y., & Courville, A. (2016). Deep learning. MIT Press.
- Manyika, J., Chui, M., Farrell, D., Stamenov, K., & Chaturvedi, R. (2019). Notes from the AI frontier: Applications and value of deep learning. McKinsey Global Institute.

---

[i] LaPorte, T. R., & Consolini, P. M. (1991). Working in practice but not in theory: Theoretical challenges of high-reliability organizations. Journal of Public Administration Research and Theory, 1(1), 19–48

- [ii] Siswanto, & Yuliana, I. (2022). Linking transformational leadership with job satisfaction: The mediating roles of trust and team cohesiveness. The Journal of Management Development, 41(2), 94–117. https://doi.org/10.1108/JMD-09-2020-0293
- [iii] Perrow, C. (2011). Normal accidents: Living with high-risk technologies (Updated ed.). Princeton University Press

- [iv] Reason, J. (2016). Managing the risks of organizational accidents. Routledge.
- [v] Weick et al., 2008
- [vi] LaPorte, T. R., & Consolini, P. M. (1991). Working in practice but not in theory: Theoretical challenges of high-reliability organizations. Journal of Public Administration Research and Theory, 1(1), 19–48.
- [vii] Weick, K. E., & Roberts, K. H. (1993). Collective mind in organizations: Heedful interrelating on flight decks. Administrative Science Quarterly, 38(3), 357–381. https://doi.org/10.2307/2393372
- [viii] Perrow, C. (2011). Normal accidents: Living with high-risk technologies (Updated ed.). Princeton University Press.
- [ix] Weick, K. E., Sutcliffe, K. M., & Obstfeld, D. (1999). Organizing for high reliability: Processes of collective mindfulness. In R. S. Sutton & B. M. Staw (Eds.), Research in organizational behavior (Vol. 21, pp. 81–123). JAI Press.
- [x] Van Knippenberg, D., & Stam, D. (2014). Visionary leadership. In D. V. Day (Ed.), The Oxford handbook of leadership and organizations (pp. 241–259). Oxford University Press
- [xi] Rosso, B. D., Dekas, K. H., & Wrzesniewski, A. (2010). On the meaning of work: A theoretical integration and review. Research in Organizational Behavior, 30, 91–127
- [xii] Leong, M. K., Mehmood, K., Ishrat, K., Ali, H., & Sadiq, M. (2025). Grandiose narcissism, leadership, and workplace deviance: Does self-interest always breed toxicity in leadership roles? BMC Psychology, 13(1), 963–16. https://doi.org/10.1186/s40359-025-03301-y
- [xiii] Harms, P. D., Bourgoin, A., Silard, A., & Wright, S. (2025). The lonely facade of narcissistic CEOs: Surface acting among narcissistic CEOs in response to feelings of loneliness. Personality and Individual Differences, 241, 113186
- [xiv] Martin, S. R., Emich, K. J., McClean, E. J., & Woodruff, T. (2022). Keeping teams together: How ethical leadership moderates the effects of performance on team efficacy and social integration. Journal of Business Ethics, 176(1), 127–139. https://doi.org/10.1007/s10551-020-04685-0

www.ingramcontent.com/pod-product-compliance
Lightning Source LLC
LaVergne TN
LVHW051134210926
848482LV00006B/309